Rho GTPases
Molecular Biology in Health and Disease

Rho GTPases
Molecular Biology in Health and Disease

editors

Philippe Fort
French National Center for Scientific Research (CNRS), France

Anne Blangy
French National Center for Scientific Research (CNRS), France

World Scientific

NEW JERSEY · LONDON · SINGAPORE · BEIJING · SHANGHAI · HONG KONG · TAIPEI · CHENNAI · TOKYO

Published by

World Scientific Publishing Co. Pte. Ltd.
5 Toh Tuck Link, Singapore 596224
USA office: 27 Warren Street, Suite 401-402, Hackensack, NJ 07601
UK office: 57 Shelton Street, Covent Garden, London WC2H 9HE

Library of Congress Cataloging-in-Publication Data
Names: Fort, Philippe, editor. | Blangy, Anne, editor.
Title: Rho GTPases : molecular biology in health and disease / [edited by]
    Philippe Fort, Anne Blangy.
Other titles: Rho GTPases (Fort)
Description: New Jersey : World Scientific, 2017. | Includes bibliographical references and index.
Identifiers: LCCN 2017026747 | ISBN 9789813228788 (hardcover : alk. paper)
Subjects: | MESH: rho GTP-Binding Proteins | Signal Transduction
Classification: LCC QP551 | NLM QU 55.2 | DDC 572/.633--dc23
LC record available at https://lccn.loc.gov/2017026747

British Library Cataloguing-in-Publication Data
A catalogue record for this book is available from the British Library.

For any available supplementary material, please visit
http://www.worldscientific.com/worldscibooks/10.1142/10674#t=suppl

Typeset by Stallion Press
Email: enquiries@stallionpress.com

Printed in Singapore

Preface

Rho GTPase: *Molecular Biologyin Health and Disease* is an updated review on the biology of Rho GTPases and constitutes an essential reading for molecular and cell biologists. It is also an invaluable guide to post-graduate and medical students who wish to catch up or deepen their knowledge in cell biology.

GTPases of the Ras homologs (Rho) family were discovered in the early 90s as major regulators of cytoskeletal remodeling. Like most Ras-like members, they behave as signaling relays, switched on when bound to GTP, off when bound to GDP. Already present at the origin of eukaryotes, Rho GTPases control many aspects of cell physiology: polarity, endo/exocytosis, adhesion, motility, transcriptional activation, cell cycle progression, or apoptosis, among others. In view of such pleiotropic activities, Rho-controlled pathways have proven to be of medical relevance, in particular in tumorigenesis, hypertension, bone loss, or infectious diseases.

Rho GTPase: *Molecular Biology in Health and Disease* provides readers the basic knowledge in understanding how Rho family members behave biochemically and how this influences cell properties. The book is divided into three parts.

Part 1 concerns the basic Rho signaling module, i.e., Rho GTPases and their regulators. Chapter 1 gives an evolutionary perspective of Rho family repertoires, in particular how complexity has increased in multicellular organisms. Chapter 2 specifically reviews atypical members, which are not regulated as the canonical Rac, Cdc42, and RhoA are. Half of Rho GTPases in vertebrates are atypical and control particular cell features. Chapter 3 gives an overview of the many regulators of typical Rho GTPases. In human, over 140 regulators have been identified and understanding which regulators control which GTPases and how regulators get themselves activated remains a daunting challenge.

Part 2 addresses the fundamental aspects of multicellularity in which Rho pathways play key roles. Chapter 4 describes how controlling the formation of epithelial junctions, contact inhibition of locomotion, and cell fusion is pivotal for embryonic development and tissue homeostasis and how their deregulation influences the progression of several pathologies. Essential too in these processes is mechanotransduction. Chapter 5 reviews how multicellularity generates mechanical stress between cells, and how this constitutes signals that Rho-controlled pathways convert into cytoskeletal rearrangements. Chapter 6 addresses how infectious bacteria can switch off Rho signaling by means of posttranslational modifications. This neutralizes the host's immune system and reduces epithelial and endothelial barriers, thereby favoringthe entry of bacteria.

Part 3 describes several pathophysiological processes controlled by Rho signaling. Chapter 7 describes how mutations may classify *Rho* genes as oncogenes or tumor suppressors depending on cancer types. Chapter 8 shows how complex the interplay between several Rho GTPases to control osteoclast differentiation and bone remodeling is. Chapter 9 gives a successful example of translational research, from the specific inhibition of a positive regulator of Rac to the development of new molecules against osteoporosis. Last, Chapter 10 describes how leukocytes travel across endothelia lining blood vessels, a complex process that relies on a series of Rho-dependent mechanisms.

We editors wish to dedicate this review to the memory of Alan Hall and Chris Marshall. We deeply miss these two pioneers in the Ras-like GTPases field, who had such a great influence on the cell biology community.

Anne Blangy,
Philippe Fort

CRBM, CNRS-UMR5237
Montpellier, France

Contents

Part 1
Rho signaling components

Rho signaling: An historical and evolutionary perspective

1

Philippe Fort*

*Centre de Recherche en Biologie cellulaire de Montpellier (CRBM),
CNRS-UMR5237, Université de Montpellier, 1919 route de Mende
34293 Montpellier cedex 05, France*
*philippe.fort@crbm.cnrs.fr

Keywords: Rho GTPases, Ras-like, signaling, evolution.

1.1. Introduction

Ras homologs (Rho) GTPases are low molecular weight proteins (21–28 kDa). They form with Arf, Rab, Ran and Ras a distinct family of the Ras-like superfamily. Like other Ras-like GTPases, Rho GTPases act as biochemical gates: they are inactive when bound to GDP and active when bound to GTP. Once activated, they acquire the ability to bind effectors that mediate the cellular response (Bourne *et al.*, 1991). Three types of regulators control Rho activity (Geyer and Wittinghofer, 1997): Guanine nucleotide exchange factors (GEF), which promote the conversion to the GTP-bound form; GTPase activating proteins (GAP), which stimulate the intrinsic GTPase activity of the Rho protein thereby promoting the conversion to the GDP-bound form; Guanine dissociation inhibitors (GDI), which control access of Rho GTPases to other regulators, effectors and membranes (see Chapter 3 by Amin and Ahmadian).

Ras-like proteins share in common a structural unit responsible for nucleotide binding and hydrolysis (G1–5 boxes, Fig. 1) and two regions that specifically bind

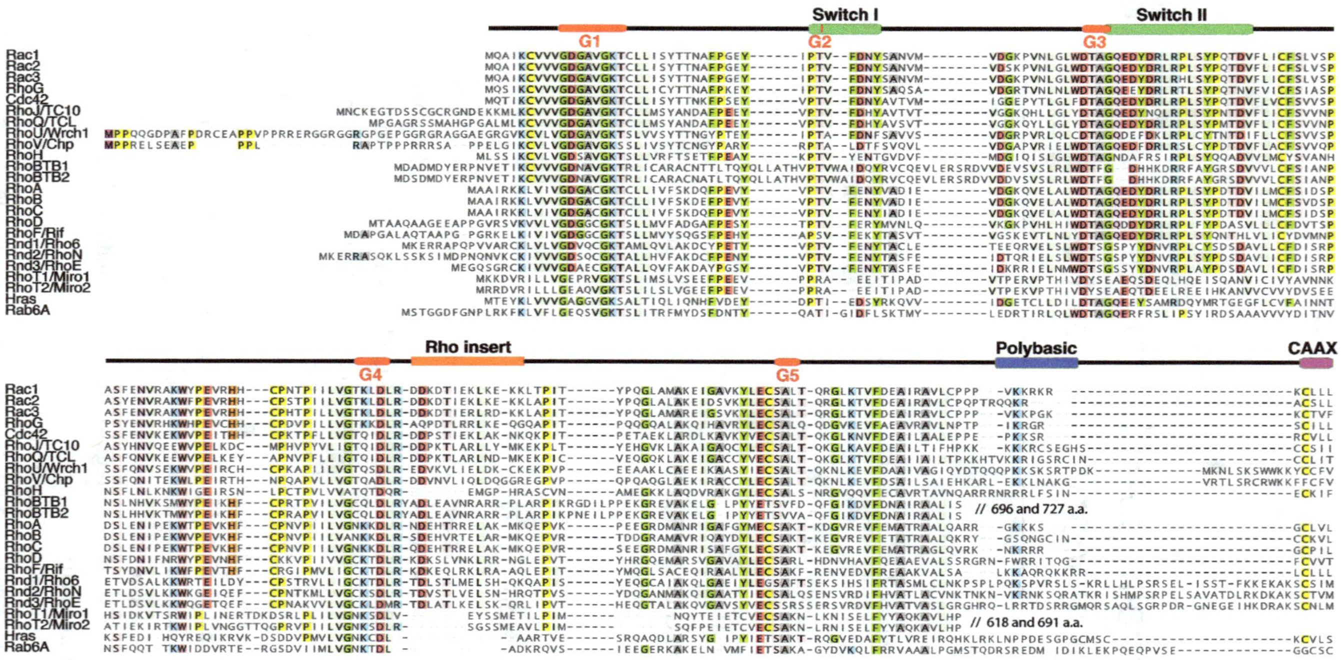

Figure 1. Structural features of human Rho proteins and their conservation across the family. G1–G5: Guanine nucleotide binding boxes. RhoT/Miro, Ha-Ras and Rab6A represent outgroup sequences. RhoBTB protein sequences (696 and 727 amino acids) and Miro (618 and 691 amino acids) were truncated.

regulators or effectors, the switch 1 and 2 regions (Bourne *et al.*, 1990). The two switch regions are highly conserved among all Ras-like GTPases and show dramatic conformation changes between the GTP- and GDP-bound states responsible for the differential binding to effectors. Rho GTPases also have C-terminal CAAX motifs (C: cysteine, A: aliphatic, X: any amino acid), which undergo post-translational lipid modifications responsible for membrane targeting, and a polybasic region adjacent to the CAAX motif, which contributes to association with membranes, interaction with regulators and subcellular localization (Williams, 2003). Rho GTPases also share a unique feature, the presence of a 10–15 residue "Rho insert", important for the regulation of their activity (Hakoshima *et al.*, 2003).

Four years later after the discovery of the first Rho GTPases (Madaule and Axel, 1985), the link between Rho activity and the dynamics of the actin cytoskeleton was inferred from the observation that *C. botulinum* C3 toxin treatment led concomitantly to ADP-ribosylation of RhoC and to the disappearance of F-actin microfilaments in Vero cells (Chardin *et al.*, 1989). The causal role of Rho activity on cytoskeletal organization was formally demonstrated by the pioneering work from Alan Hall's lab (Paterson *et al.*, 1990; Ridley and Hall, 1992; Ridley *et al.*, 1992). By microinjecting recombinant active or dominant negative mutated RhoA and Rac1 proteins, they established that Rac1 controls cell spreading through membrane ruffling while RhoA controls cell contraction through F-actin stress fiber bundling. They showed later on that Cdc42 controls the formation of filopodia, a third type of F-actin structure also involved in cell motility and spreading (Nobes and Hall, 1995). They also showed that Rac1 and RhoA are activated by growth factors (Nobes *et al.*, 1995).

The three types of regulators were identified during the same period of time. A screen for oncogenic DNA in NIH3T3 cells identified Dbl, cloned from a human diffuse B-cell lymphoma (Eva and Aaronson, 1985). Dbl was later shown to be a GEF active on the Cdc42 protein (Hart *et al.*, 1991; Ron *et al.*, 1991). In 2002, two teams identified a second RhoGEF family, the DOCK/CZH family, unrelated to Dbl RhoGEFs and active on Cdc42 and Rac (Côté and Vuori, 2002; Meller *et al.*, 2002). Besides, RhoGAP activities toward RhoA and Rac were detected in vertebrate cell extracts (Garrett *et al.*, 1989), among which p50RhoGAP, n-chimaerin and Bcr were identified (Garrett *et al.*, 1991). Bcr is a gene frequently rearranged in chronic myeloid lymphomas (Groffen *et al.*, 1984; Heisterkamp *et al.*, 1985; Shtivelman *et al.*, 1985). Last, two RhoGDI were characterized from brain extracts (Fukumoto *et al.*, 1990; Matsui *et al.*, 1990).

The functional link between Rho signaling pathways, cell migration and oncogenesis initiated a highly active field of research that has since then generated over

11,500 publications. It emerges from these numerous studies that Rho signaling pathways participate in the control of a wide variety of fundamental cell properties, such as adhesion and motility, proliferation, differentiation and apoptosis. For this reason, this is no surprise that they are implicated in physiological processes, be it normal, like development or the immune response, or pathological, like hypertension or cancer.

This chapter gives a historical review on the discovery of the mammalian Rho GTPase repertoire and an evolutionary perspective of its ontogeny.

1.2. Identification of the human repertoire of Rho GTPases

Most sequences for Rho GTPases were identified by molecular biology techniques based on low stringency DNA hybridization. At this time, sequence determination was a time and labor consuming process, and genomes completely sequenced were only viral and mitochondrial ones.

The first Rho member was isolated serendipitously in 1985 from a low stringency screen of a Californian sea hare (*Aplysia californica*) cDNA library (Madaule and Axel, 1985). The deduced 192 amino acid protein shared 35% similarity with H-Ras and was thus termed Rho, as they were the first Ras homologs identified. Homologs to the *Aplysia* Rho gene were next found in human (RhoA, RhoB and RhoC, Chardin *et al.*, 1988; Yeramian *et al.*, 1987) and in the yeast *Saccharomyces cerevisiae* (RHO-1 and RHO-2, Madaule *et al.*, 1987).

The use of degenerate oligonucleotides corresponding to amino acid motifs conserved in Ras-like GTPases led to the identification of the following: the human Rac1 and Rac2 (for Ras related C3 botulinum toxin substrate) (Didsbury *et al.*, 1989), which actually turned out later to be poor C3 toxin substrates (Ménard *et al.*, 1992); the yeast and human CDC42 (cell division cycle) proteins (Johnson and Pringle, 1990; Munemitsu *et al.*, 1990; Shinjo *et al.*, 1990); RhoJ/TC10, closely related to CDC42 (Drivas *et al.*, 1990); RhoD (Chavrier *et al.*, 1992); Rac3 (Courjal *et al.*, 1997; Haataja *et al.*, 1997) and three Rnd proteins (Foster *et al.*, 1996; Nobes *et al.*, 1998). RhoG, a distant relative to Rac proteins, was picked from a differential screening for growth-induced genes (Vincent *et al.*, 1992). RhoH/TTF was identified from a translocation event frequently observed in B cell lymphomas (Dallery *et al.*, 1995). The two closely related RhoV/Chp (Cdc42 homologous protein) and RhoU/Wrch1 (Wnt-1 responsive Cdc42 homolog) were identified through the ability to bind p21 activated kinases (PAK) (RhoV, Aronheim *et al.*, 1998) and to be induced by Wnt signaling (RhoU, Tao *et al.*, 2001).

The availability of complete mammalian genomes and EST databases allowed to identify the missing members: RhoQ/TCL (TC10-like, Vignal *et al.*, 2000),

RhoF/Rif (Rho in filopodia, Ellis and Mellor, 2000) and the two RhoBTB proteins, distantly related to other Rho members and homologs to the unique RhoBTB present in the genome of the amoeba *Dictyostelium discoideum* (Rivero *et al.*, 2001). RhoBTB are 600–700 amino acids long and have a N-terminal Rho GTPase domain fused to a Broad-complex, Tramtrack and Bric a brac (BTB) protein–protein interaction domain. The two mitochondrial Miro proteins, at first grouped into the Rho family because of the similarity of their N-terminal GTP-binding domains (Fransson *et al.*, 2003), were later on considered as a closely related but likely distinct Ras-like family (Boureux *et al.*, 2007).

The Rho family in human thus comprises 18 members with a simple Ras-like structure and two members, the RhoBTBs, in which the Ras-like domain is fused to a second functional domain. The 20 members are grouped into two major clusters (Rac-like and Rho-like).

1.3. Ontogeny and early evolution of Rho GTPases

Like other multigene family members, Rho GTPases are derived from a common ancestor gene present in an extinct ancestral organism. The amino acid similarity among the 20 human Rho members ranges from 51% to 98%, indicating they experienced differential selective pressures and/or emerged at different times.

Figure 2 shows the relationships between human Rho family members using Maximum Likelihood and Bayesian probabilistic analyses. The 20 sequences are distributed between two groups: The Rho group includes RhoABC, RhoDF and Rnd123; the Rac group includes Cdc42/RhoJQ, RhoUV, Rac123/RhoG, RhoH and RhoBTB. Note that although moderately supported, the clustering of RhoDF and Rnd123 is in agreement with the identical exon/intron structure of the five genes. Also, note that RhoG is not clustered with Rac, which is not the case when using a distance-based tree inference (Boureux *et al.*, 2007).

Reconstructing the evolutionary events that led to current Rho families in extant species can help solve this type of inconsistencies. To achieve this, it was necessary (i) to determine the full Rho GTPases repertoires in key phylogenetic groups and (ii) to examine the orthology and paralogy status of members across repertoires.

These notions are critical for evolutionary studies (Sonnhammer and Koonin, 2002): orthologs are genes from different species that derive from a single gene in their last common ancestor; paralogs are genes that derive from duplications of a single gene in a genome, like all Rho members in the Fig. 2. They include inparalogs, when duplications are specific of a lineage, and outparalogs, when duplications occurred before the divergence of several lineages. These concepts are needed for

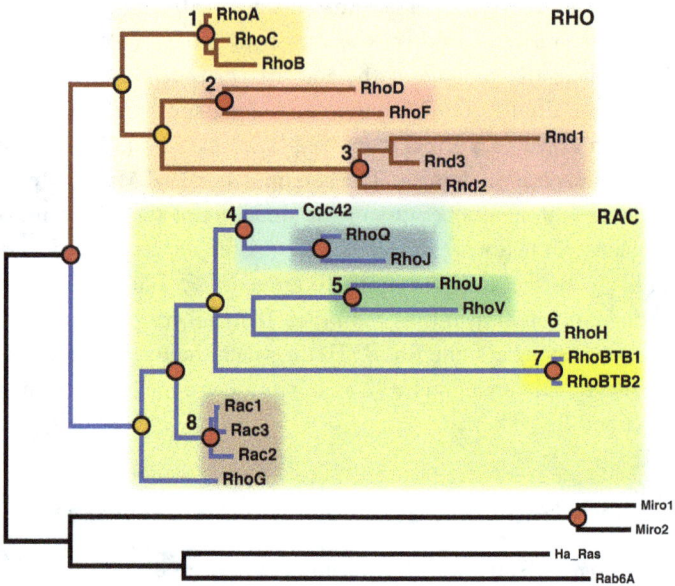

Figure 2. Phylogenetic relationships between human Rho members. Maximum Likelihood and MrBayes approaches were used to infer the tree. Red circles indicate nodes strongly supported by both analyses; yellow circles, strongly supported by only one method.

deducing functional similarities from studies in distinct biological models. Indeed, in most cases orthologs are expected to fulfil similar cellular functions, while paralogs should differ by one or several features (Katju and Bergthorsson, 2013). Note that most gene duplications in an organism are inherently unstable and probably detrimental to the fitness. Except the case where a duplication is retained because environmental changes exert a selective pressure for more of the same product, the fixation and maintenance of duplicated copies over generations are associated with accumulation of mutations in the extra copy. Thus, paralogs are likely submitted to different evolution rates, as long as the extra copy has not gained a new function (Katju, 2012).

All these considerations are particularly important for Rho families, which have evolved mainly by independent gene duplication events and have generated many in- and out-paralogs.

As mentioned above, Rho family members were first identified in mollusks (*A. californica*), fungi (*S. cerevisiae*) and mammals (*H. sapiens, M. musculus*). This suggested a very ancient origin, since Fungi and Metazoa belong to the Opisthokonta branch that diverged 1–1.2 billion years ago (Parfrey *et al.*, 2011).

During the last decade, whole genome sequences projects gave access to species of particular evolutionary interest and allowed the identification of Rho GTPases repertoires of fishes (Salas-Vidal *et al.*, 2005), tunicates (Philips *et al.*, 2003) and amoebozoa (Rivero *et al.*, 2001).

Two global analyses (Boureux *et al.*, 2007; Eliáš and Klimeš, 2012), covering most eukaryotic clades, led to the following conclusions:

(i) The phylogenetic relationships between Rho GTPases cannot be strongly supported: Rho GTPases are small proteins with a globular structure. Sequence can be aligned along 170 amino acids, including 42 amino acids involved in the binding to the nucleotide and effectors and thus highly constrained. The number of informative sites is thus reduced and the evolution rate is unequally distributed along the molecule. This precludes accurate and statistically supported inference of phylogenetic relationships among Rho genes on a long evolutionary scale. However, in most cases, Rho GTPases can confidently be assigned as relatives to Rac or Rho s.s. (*stricto sensu*, i.e., the ancestor to human RhoA-C).

(ii) Rho GTPases were present in the last eukaryotic common ancestor (LECA), i.e., 1.4–1.8 billion years ago (Parfrey *et al.*, 2011). This is concluded from the occurrence of Rho genes in all eukaryotic supergroups examined. Since Rho GTPases in non-Opisthokont lineages are more closely related to Rac proteins, Rac is likely the founder member of the whole family. The fact that Rho genes are absent in sub-clades of Chlorophyta, Trypanosomatidae, Stramenopiles and Alveolates must be considered as multiple independent loss events (Fig. 3).

(iii) RhoBTB are found in Metazoa, Amoebozoa, Apusozoa, Filasterea and Heterolobosea. Although the exact position of the eukaryotic root as well as positions of a few specific lineages are still debated, the phylogenetic relationships between major eukaryotic lineages are well established (Adl *et al.*, 2012; Eme *et al.*, 2014). The first four clades mentioned above belong to Amorphea, which also include Fungi, Ichthyosporea and Choanoflagellida (Fig. 3). Since the three latter clades do not have RhoBTB genes, this implies that RhoBTB was present in the last ancestor of Amorphea and was then lost independently at least three times.

The events leading to the presence of RhoBTB in Heterolobosea but not in other excavates cannot be established with confidence, since this greatly depends on the eukaryotic root position (Fig. 3). Whatever its precise ontogeny, RhoBTB is nevertheless one of the eldest Rac paralog.

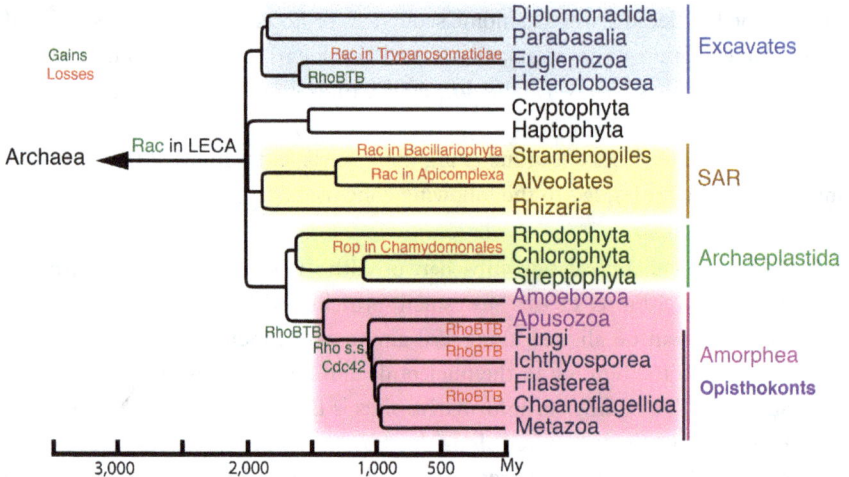

Figure 3. Ontogeny of Rho GTPases in early eukaryotes. Shown is the phylogeny of eukaryotic supergroups as published by Adl *et al.* (2012) along with a timescale in million years (My). Average divergence times are derived from Hedges *et al.* (2006). LECA: last eukaryotic common ancestor. Archaeplastida encode Rac-like proteins termed as Rho of plants (Rop, Yang and Watson, 1993).

(iv) Rho s.s. is detected only in Opisthokonts and Apusozoa, while Cdc42 is detected only in Opisthokonts. This establishes that Rac, Rho s.s. and Cdc42 were present at least at the onset of Opisthokonts. Apusozoa and Amoebozoa samples are too sparse to allow firm conclusions on when Rho s.s. and Cdc42 duplicated first. The Opisthokont sub-clades Fungi, Choanoflagellates and Filasteria have qualitatively comparable repertoires, made of a single Cdc42 and at least one Rac and one Rho s.s.

1.4. Diversification of the Rho family in Metazoa

Porifera (sponges) is the basalmost metazoan clade and represents over 5,000 species. The sponges *Suberites domuncula* and *Amphimedon queenslandica* have a simple Rho repertoire, made of one Rac, three Rho s.s. and one Cdc42 (Fig. 4). The repertoire is very similar to those of Choanoflagellida (*Monosiga brevicollis, Salpingoeca rosetta*), Ichthyosporea (*Sphaeroforma arctica*) and Filasterea (*Capsaspora owczarzaki*).

Rnd and RhoUV proteins emerged in Cnidaria: Rnd genes were identified in several genomes of Anthozoa (the sea anemones *Nematostella vectensis, Acropora digitifera* and *Exaiptasia pallida*) and Hydrozoa (*Hydra vulgaris*); RhoUV genes are

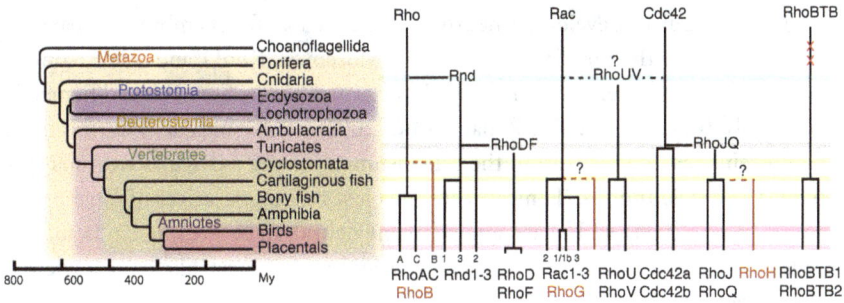

Figure 4. Ontogeny of the mammalian Rho repertoire. Shown is the phylogeny of Metazoa as published by Adl *et al.* (2012) along with a timescale in million years (My). Average divergence times are derived from Hedges *et al.* (2006).

present in *N. vectensis* and *E. pallida* genomes. Cnidaria have also one Rac, one Cdc42, four Rho s.s. and two RhoBTB. Rnd and RhoUV are atypical in that they have substitutions in their G3 box. Consequently, they are devoid of GTPase activity and remain active until they are degraded (see Chapter 2 by Aspenström). In mammals, Rnd proteins antagonize RhoA-dependent processes and have pleiotropic effects (Chardin, 2006). RhoU and RhoV control cell adhesion and migration in cultured cells and developmental processes (Aspenström *et al.*, 2007; Fort and Théveneau, 2014). In contrast to Porifera, which lack typical muscles and neurons, cnidarians share with bilaterians smooth and striated muscles. Cnidarians also possess a neural net, and some of them, nerve cords and plexuses (Lanna, 2015). In mice, Rnd3 deficiency elicits neuromuscular defects with a reduced number of motor neurons (Mocholí *et al.*, 2011) and RhoU is involved in heart development and failure (Dickover *et al.*, 2014; Herrer *et al.*, 2014). The use of a Cnidaria model is needed to address if functions of the ancestral Rnd and RhoUV GTPases also pertained to nerve and muscle biology.

Three features were added to the Rho repertoire in Tunicates: The selection of an alternative C-terminus for Cdc42 and the appearance of RhoDF and RhoJQ. Cdc42 alternative splicing has long remained neglected until recent works showed striking differences in the ability of the two isoforms to induce filopodia (Wirth *et al.*, 2013). The ubiquitous Cdc42 isoform is prenylated, whereas the alternative isoform is brain specific and can be palmitoylated thanks to the cysteine doublet of the CAAx box. In neuroblastoma cells, lipidation of the two C-terminal cysteines is essential for proper membrane binding of the isoform and establishment of its activities, in particular the induction of densely packed filopodia. Brain specific Cdc42 isoform is mostly expressed in the hippocampus, where it plays critical roles in the formation of dendritic filopodia and spines. Besides,

RhoQ is essential in developing neurons, in particular for membrane expansion during axon growth (Gracias *et al.*, 2014). Moreover, RhoD and RhoF control neurite retraction in cells stimulated with the semaphorins Sema3A and Sema6A, respectively (Fan *et al.*, 2015; Zanata *et al.*, 2002). This suggests that the three Rho proteins may participate in the establishment of the particular nervous system of Tunicates. Indeed, Tunicates are the closest living relatives of Vertebrates (Delsuc *et al.*, 2006) and share among them neural structures, such as mid-to hindbrain boundary, placodes, motoneurons and neural crest-like cells (reviewed in Lemaire, 2011).

Many Rho members were generated between Tunicates and bony fishes. Two whole genome duplications (WGDs) occurred before the Cyclostomata (lampreys) and Gnathostomata (jawed vertebrates) (Smith *et al.*, 2013). However, most genes for new Rho members are present in cartilaginous fishes and not in lampreys (Fig. 4). Since no WGDs occurred between these two clades, this suggests that either the genes have been missed in current assemblies of lamprey genomes or they were secondary lost in Cyclostomata.

Three new genes, RhoB, RhoG and RhoH, appeared by retrotransposition and not by gene duplication, as concluded from the monoexonic structure of their coding sequences. RhoB and RhoG were likely derived from RhoAC and Rac1-3, respectively. The origin of RhoH remains murkier, although probabilistic phylogenetic analyses suggest a branching with the Rac/Cdc42/RhoUV/RhoBTB subgroup (Fig. 2). This is supported by the presence of synapomorphic sites shared by all these members. As shown in Fig. 4, the mammalian Rho family was nearly shaped at the onset of bony vertebrates. Only two members were not yet present: Rac1b, an alternatively spliced Rac1 isoform, and RhoD, a duplicated from RhoF.

RhoD was only found in therian genomes. Its evolution rate was 4.4 times faster than that of RhoF, indicating a relaxed selective pressure (Boureux *et al.*, 2007). As mentioned above, RhoD controls endosome dynamics (Murphy *et al.*, 1996) and plays roles in axon guidance (Zanata *et al.*, 2002). The physiological relevance of this function in Amniotes remains to be clarified.

Rac1b has a 19 a.a. insertion resulting from inclusion of an alternatively spliced 57 bp exon cassette (Jordan *et al.*, 1999). The result of this insertion is a higher activity and a selective downstream signaling. Exons encoding the 19 a.a. peptide were found exclusively in the genomes of Amniotes (Boureux *et al.*, 2007). Although reports have documented the role of Rac1b in the progression of many cancer types (listed in Faria *et al.*, 2016), the normal function of this isoform remains unknown.

1.5. Concluding remarks

The aim of this chapter is to give an insight into the evolution processes that shaped the current Rho repertoire in mammals and, in particular, in humans. Inparalogs found in other clades have not been mentioned since their functions are beyond the scope of this book, focused on Rho signaling in health and disease. An example is the Rac-like Mtl/Mig2, specifically found in Ecdysozoa, which controls axon guidance in *C. elegans* and *D. melanogaster* (Ng *et al.*, 2002; Xu *et al.*, 2015).

The current evolutionary picture shows that Rac is the founder member of the Rho family, which gave rise to Rho s.s., Cdc42 and RhoBTB before Metazoa. This is in agreement with the key roles of the Rac, Rho s.s. and Cdc42 in basic cell biology and suggests it might also be the case for RhoBTB. The timing of emergence of Rnd and RhoUV is consistent with their roles in the acquisition of muscle and nerve cells, and for Cdc42 isoforms, RhoJQ and RhoDF, in formation of the vertebrate central nervous system.

Many members or isoforms have partially known biological functions and recent literature mostly pertains to their implication in pathologies, mainly cancer. Specific biological models are needed to understand which processes they control in normal conditions.

Evolutionary analyses are currently being run on RhoGEF and RhoGAP families, which represent more than 150 proteins in human. This will help classify these big families and allow to identify which RhoGEF, RhoGTPase and RhoGAP have coevolved and constitute coregulated modules.

References

Adl, S.M., Simpson, A.G.B., Lane, C.E., Lukeš, J., Bass, D., Bowser, S.S., Brown, M.W., Burki, F., Dunthorn, M., Hampl, V., *et al.* (2012). The revised classification of eukaryotes. J. Eukaryot. Microbiol. *59*, 429–514.

Aronheim, A., Broder, Y.C., Cohen, A., Fritsch, A., Belisle, B., and Abo, A. (1998). Chp, a homologue of the GTPase Cdc42Hs, activates the JNK pathway and is implicated in reorganizing the actin cytoskeleton. Curr. Biol. *8*, 1125–1129.

Aspenström, P., Ruusala, A., and Pacholsky, D. (2007). Taking Rho GTPases to the next level: the cellular functions of atypical Rho GTPases. Exp. Cell Res. *313*, 3673–3679.

Boureux, A., Vignal, E., Faure, S., and Fort, P. (2007). Evolution of the Rho family of ras-like GTPases in eukaryotes. Mol. Biol. Evol. *24*, 203–216.

Bourne, H.R., Sanders, D.A., and McCormick, F. (1990). The GTPase superfamily: a conserved switch for diverse cell functions. Nature *348*, 125–132.

Bourne, H.R., Sanders, D.A., and McCormick, F. (1991). The GTPase superfamily: conserved structure and molecular mechanism. Nature *349*, 117–127.

Chardin, P. (2006). Function and regulation of Rnd proteins. Nat. Rev. Mol. Cell Biol. *7*, 54–62.

Chardin, P., Madaule, P., and Tavitian, A. (1988). Coding sequence of human rho cDNAs clone 6 and clone 9. Nucleic Acids Res. *16*, 2717.

Chardin, P., Boquet, P., Madaule, P., Popoff, M.R., Rubin, E.J., and Gill, D.M. (1989). The mammalian G protein rhoC is ADP-ribosylated by *Clostridium botulinum* exoenzyme C3 and affects actin microfilaments in Vero cells. EMBO J. *8*, 1087–1092.

Chavrier, P., Simons, K., and Zerial, M. (1992). The complexity of the Rab and Rho GTP-binding protein subfamilies revealed by a PCR cloning approach. Gene *112*, 261–264.

Côté, J.-F. and Vuori, K. (2002). Identification of an evolutionarily conserved superfamily of DOCK180-related proteins with guanine nucleotide exchange activity. J. Cell Sci. *115*, 4901–4913.

Courjal, F., Chuchana, P., Theillet, C., and Fort, P. (1997). Structure and chromosomal assignment to 22q12 and 17qter of the ras-related Rac2 and Rac3 human genes. Genomics *44*, 242–246.

Dallery, E., Galiègue-Zouitina, S., Collyn-d'Hooghe, M., Quief, S., Denis, C., Hildebrand, M.P., Lantoine, D., Deweindt, C., Tilly, H., and Bastard, C. (1995). TTF, a gene encoding a novel small G protein, fuses to the lymphoma-associated LAZ3 gene by t(3;4) chromosomal translocation. Oncogene *10*, 2171–2178.

Delsuc, F., Brinkmann, H., Chourrout, D., and Philippe, H. (2006). Tunicates and not cephalochordates are the closest living relatives of vertebrates. Nature *439*, 965–968.

Dickover, M., Hegarty, J.M., Ly, K., Lopez, D., Yang, H., Zhang, R., Tedeschi, N., Hsiai, T.K., and Chi, N.C. (2014). The atypical Rho GTPase, RhoU, regulates cell-adhesion molecules during cardiac morphogenesis. Dev. Biol. *389*, 182–191.

Didsbury, J., Weber, R.F., Bokoch, G.M., Evans, T., and Snyderman, R. (1989). rac, a novel ras-related family of proteins that are botulinum toxin substrates. J. Biol. Chem. *264*, 16378–16382.

Drivas, G.T., Shih, A., Coutavas, E., Rush, M.G., and D'Eustachio, P. (1990). Characterization of four novel ras-like genes expressed in a human teratocarcinoma cell line. Mol. Cell. Biol. *10*, 1793–1798.

Eliáš, M. and Klimeš, V. (2012). Rho GTPases: deciphering the evolutionary history of a complex protein family. Methods Mol. Biol. *827*, 13–34.

Ellis, S. and Mellor, H. (2000). The novel Rho-family GTPase Rif regulates coordinated actin-based membrane rearrangements. Curr. Biol. *10*, 1387–1390.

Eme, L., Sharpe, S.C., Brown, M.W., and Roger, A.J. (2014). On the age of eukaryotes: evaluating evidence from fossils and molecular clocks. Cold Spring Harb. Perspect. Biol. *6*, a016139.

Eva, A. and Aaronson, S.A. (1985). Isolation of a new human oncogene from a diffuse B-cell lymphoma. Nature *316*, 273–275.

Fan, L., Yan, H., Pellegrin, S., Morigen, and Mellor, H. (2015). The Rif GTPase regulates cytoskeletal signaling from plexinA4 to promote neurite retraction. Neurosci. Lett. *590*, 178–183.

Faria, M., rcia, Capinha, L., Simões-Pereira, J., Bugalho, M.J., and Silva, A.L. (2016). Extending the impact of RAC1b overexpression to follicular thyroid carcinomas. Int. J. Endocrinol. *2016*, e1972367.

Fort, P. and Théveneau, E. (2014). PleiotRHOpic: Rho pathways are essential for all stages of neural crest development. Small GTPases *5*, e27975.

Foster, R., Hu, K.-Q., Lu, Y., Nolan, K.M., Thissen, J., and Settleman, J. (1996). Identification of a novel human Rho protein with unusual properties: GTPase deficiency and *in vivo* farnesylation. Mol. Cell. Biol. *16*, 2689–2699.

Fransson, Å., Ruusala, A., and Aspenström, P. (2003). Atypical Rho GTPases have roles in mitochondrial homeostasis and apoptosis. J. Biol. Chem. *278*, 6495–6502.

Fukumoto, Y., Kaibuchi, K., Hori, Y., Fujioka, H., Araki, S., Ueda, T., Kikuchi, A., and Takai, Y. (1990). Molecular cloning and characterization of a novel type of regulatory protein (GDI) for the rho proteins, ras p21-like small GTP-binding proteins. Oncogene *5*, 1321–1328.

Garrett, M.D., Self, A.J., van Oers, C., and Hall, A. (1989). Identification of distinct cytoplasmic targets for ras/R-ras and rho regulatory proteins. J. Biol. Chem. *264*, 10–13.

Garrett, M.D., Major, G.N., Totty, N., and Hall, A. (1991). Purification and N-terminal sequence of the p21rho GTPase-activating protein, rho GAP. Biochem. J. *276 (Pt 3)*, 833–836.

Geyer, M. and Wittinghofer, A. (1997). GEFs, GAPs, GDIs and effectors: taking a closer (3D) look at the regulation of Ras-related GTP-binding proteins. Curr. Opin. Struct. Biol. *7*, 786–792.

Gracias, N.G., Shirkey-Son, N.J., and Hengst, U. (2014). Local translation of TC10 is required for membrane expansion during axon outgrowth. Nat. Commun. *5*, 3506.

Groffen, J., Stephenson, J.R., Heisterkamp, N., de Klein, A., Bartram, C.R., and Grosveld, G. (1984). Philadelphia chromosomal breakpoints are clustered within a limited region, bcr, on chromosome 22. Cell *36*, 93–99.

Haataja, L., Groffen, J., and Heisterkamp, N. (1997). Characterization of RAC3, a novel member of the Rho family. J. Biol. Chem. *272*, 20384–20388.

Hakoshima, T., Shimizu, T., and Maesaki, R. (2003). Structural basis of the Rho GTPase signaling. J. Biochem. (Tokyo) *134*, 327–331.

Hart, M.J., Eva, A., Evans, T., Aaronson, S.A., and Cerione, R.A. (1991). Catalysis of guanine nucleotide exchange on the CDC42Hs protein by the dbl oncogene product. Nature *354*, 311–314.

Hedges, S.B., Dudley, J., and Kumar, S. (2006). Time Tree: a public knowledge-base of divergence times among organisms. Bioinformatics *22*, 2971–2972.

Heisterkamp, N., Stam, K., Groffen, J., de Klein, A., and Grosveld, G. (1985). Structural organization of the bcr gene and its role in the Ph' translocation. Nature *315*, 758–761.

Herrer, I., Roselló-Lletí, E., Rivera, M., Molina-Navarro, M.M., Tarazón, E., Ortega, A., Martínez-Dolz, L., Triviño, J.C., Lago, F., González-Juanatey, J.R., et al. (2014). RNA-sequencing analysis reveals new alterations in cardiomyocyte cytoskeletal genes in patients with heart failure. Lab. Investig. J. Tech. Methods Pathol. *94*, 645–653.

Johnson, D.I. and Pringle, J.R. (1990). Molecular characterization of CDC42, a *Saccharomyces cerevisiae* gene involved in the development of cell polarity. J. Cell Biol. *111*, 143–152.

Jordan, P., Brazåo, R., Boavida, M.G., Gespach, C., and Chastre, E. (1999). Cloning of a novel human Rac1b splice variant with increased expression in colorectal tumors. Oncogene *18*, 6835–6839.

Katju, V. (2012). In with the old, in with the new: the promiscuity of the duplication process engenders diverse pathways for novel gene creation. Int. J. Evol. Biol. *2012*, 1–24.

Katju, V. and Bergthorsson, U. (2013). Copy-number changes in evolution: rates, fitness effects and adaptive significance. Front. Genet. *4*, 273.

Lanna, E. (2015). Evo-devo of non-bilaterian animals. Genet. Mol. Biol. *38*, 284.

Lemaire, P. (2011). Evolutionary crossroads in developmental biology: the tunicates. Development *138*, 2143–2152.

Madaule, P. and Axel, R. (1985). A novel ras-related gene family. Cell *41*, 31–40.

Madaule, P., Axel, R., and Myers, A.M. (1987). Characterization of two members of the rho gene family from the yeast *Saccharomyces cerevisiae*. Proc. Natl. Acad. Sci. USA *84*, 779–783.

Matsui, Y., Kikuchi, A., Araki, S., Hata, Y., Kondo, J., Teranishi, Y., and Takai, Y. (1990). Molecular cloning and characterization of a novel type of regulatory protein (GDI) for smg p25A, a ras p21-like GTP-binding protein. Mol. Cell. Biol. *10*, 4116–4122.

Meller, N., Irani-Tehrani, M., Kiosses, W.B., Del Pozo, M.A., and Schwartz, M.A. (2002). Zizimin1, a novel Cdc42 activator, reveals a new GEF domain for Rho proteins. Nat. Cell Biol. *4*, 639–647.

Ménard, L., Tomhave, E., Casey, P.J., Uhing, R.J., Snyderman, R., and Didsbury, J.R. (1992). Rac1, a low-molecular-mass GTP-binding-protein with high intrinsic GTPase activity and distinct biochemical properties. Eur. J. Biochem. *206*, 537–546.

Mocholí, E., Ballester-Lurbe, B., Arqué, G., Poch, E., Peris, B., Guerri, C., Dierssen, M., Guasch, R.M., Terrado, J., and Pérez-Roger, I. (2011). RhoE deficiency produces postnatal lethality, profound motor deficits and neurodevelopmental delay in mice. PLoS ONE *6*, e19236.

Munemitsu, S., Innis, M.A., Clark, R., McCormick, F., Ullrich, A., and Polakis, P. (1990). Molecular cloning and expression of a G25K cDNA, the human homolog of the yeast cell cycle gene CDC42. Mol. Cell. Biol. *10*, 5977–5982.

Murphy, C., Saffrich, R., Grummt, M., Gournier, H., Rybin, V., Rubino, M., Auvinen, P., Lütcke, A., Parton, R.G., and Zerial, M. (1996). Endosome dynamics regulated by a Rho protein. Nature *384*, 427–432.

Ng, J., Nardine, T., Harms, M., Tzu, J., Goldstein, A., Sun, Y., Dietzl, G., Dickson, B.J., and Luo, L. (2002). Rac GTPases control axon growth, guidance and branching. Nature *416*, 442–447.

Nobes, C.D. and Hall, A. (1995). Rho, rac, and cdc42 GTPases regulate the assembly of multimolecular focal complexes associated with actin stress fibers, lamellipodia, and filopodia. Cell *81*, 53–62.

Nobes, C.D., Hawkins, P., Stephens, L., and Hall, A. (1995). Activation of the small GTP-binding proteins rho and rac by growth factor receptors. J. Cell Sci. *108* (Pt 1), 225–233.

Nobes, C.D., Lauritzen, I., Mattei, M.G., Paris, S., Hall, A., and Chardin, P. (1998). A new member of the Rho family, Rnd1, promotes disassembly of actin filament structures and loss of cell adhesion. J. Cell Biol. *141*, 187–197.

Parfrey, L.W., Lahr, D.J., Knoll, A.H., and Katz, L.A. (2011). Estimating the timing of early eukaryotic diversification with multigene molecular clocks. Proc. Natl. Acad. Sci. USA *108*, 13624–13629.

Paterson, H.F., Self, A.J., Garrett, M.D., Just, I., Aktories, K., and Hall, A. (1990). Microinjection of recombinant p21rho induces rapid changes in cell morphology. J. Cell Biol. *111*, 1001–1007.

Philips, A., Blein, M., Robert, A., Chambon, J.-P., Baghdiguian, S., Weill, M., and Fort, P. (2003). Ascidians as a vertebrate-like model organism for physiological studies of Rho GTPase signaling. Biol. Cell *95*, 295–302.

Ridley, A.J. and Hall, A. (1992). The small GTP-binding protein rho regulates the assembly of focal adhesions and actin stress fibers in response to growth factors. Cell *70*, 389–399.

Ridley, A.J., Paterson, H.F., Johnston, C.L., Diekmann, D., and Hall, A. (1992). The small GTP-binding protein rac regulates growth factor-induced membrane ruffling. Cell *70*, 401–410.

Rivero, F., Dislich, H., Glöckner, G., and Noegel, A.A. (2001). The *Dictyostelium discoideum* family of Rho-related proteins. Nucleic Acids Res. *29*, 1068.

Ron, D., Zannini, M., Lewis, M., Wickner, R.B., Hunt, L.T., Graziani, G., Tronick, S.R., Aaronson, S.A., and Eva, A. (1991). A region of proto-dbl essential for its transforming activity shows sequence similarity to a yeast cell cycle gene, CDC24, and the human breakpoint cluster gene, bcr. New Biol. *3*, 372–379.

Salas-Vidal, E., Meijer, A.H., Cheng, X., and Spaink, H.P. (2005). Genomic annotation and expression analysis of the zebrafish Rho small GTPase family during development and bacterial infection. Genomics *86*, 25–37.

Shinjo, K., Koland, J.G., Hart, M.J., Narasimhan, V., Johnson, D.I., Evans, T., and Cerione, R.A. (1990). Molecular cloning of the gene for the human placental GTP-binding protein Gp (G25K): identification of this GTP-binding protein as the human homolog of the yeast cell-division-cycle protein CDC42. Proc. Natl. Acad. Sci. USA *87*, 9853–9857.

Shtivelman, E., Lifshitz, B., Gale, R.P., and Canaani, E. (1985). Fused transcript of abl and bcr genes in chronic myelogenous leukaemia. Nature *315*, 550–554.

Smith, J.J., Kuraku, S., Holt, C., Sauka-Spengler, T., Jiang, N., Campbell, M.S., Yandell, M.D., Manousaki, T., Meyer, A., Bloom, O.E., *et al.* (2013). Sequencing of the sea lamprey (*Petromyzon marinus*) genome provides insights into vertebrate evolution. Nat. Genet. *45*, 415–421.

Sonnhammer, E.L.L. and Koonin, E.V. (2002). Orthology, paralogy and proposed classification for paralog subtypes. Trends Genet. *18*, 619–620.

Tao, W., Pennica, D., Xu, L., Kalejta, R.F., and Levine, A.J. (2001). Wrch-1, a novel member of the Rho gene family that is regulated by Wnt-1. Genes Dev. *15*, 1796–1807.

Vignal, E., De Toledo, M., Comunale, F., Ladopoulou, A., Gauthier-Rouvière, C., Blangy, A., and Fort, P. (2000). Characterization of TCL, a new GTPase of the rho family related to TC10 and Ccdc42. J. Biol. Chem. *275*, 36457–36464.

Vincent, S., Jeanteur, P., and Fort, P. (1992). Growth-regulated expression of rhoG, a new member of the ras homolog gene family. Mol. Cell. Biol. *12*, 3138–3148.

Williams, C.L. (2003). The polybasic region of Ras and Rho family small GTPases: a regulator of protein interactions and membrane association and a site of nuclear localization signal sequences. Cell. Signal. *15*, 1071–1080.

Wirth, A., Chen-Wacker, C., Wu, Y.-W., Gorinski, N., Filippov, M.A., Pandey, G., and Ponimaskin, E. (2013). Dual lipidation of the brain-specific Cdc42 isoform regulates its functional properties. Biochem. J. *456*, 311–322.

Xu, Y., Taru, H., Jin, Y., and Quinn, C.C. (2015). SYD-1C, UNC-40 (DCC) and SAX-3 (Robo) function interdependently to promote axon guidance by regulating the MIG-2 GTPase. PLoS Genet. *11*, e1005185.

Yang, Z. and Watson, J.C. (1993). Molecular cloning and characterization of rho, a ras-related small GTP-binding protein from the garden pea. Proc. Natl. Acad. Sci. USA *90*, 8732–8736.

Yeramian, P., Chardin, P., Madaule, P., and Tavitian, A. (1987). Nucleotide sequence of human rho cDNA clone 12. Nucleic Acids Res. *15*, 1869.

Zanata, S.M., Hovatta, I., Rohm, B., and Püschel, A.W. (2002). Antagonistic effects of Rnd1 and RhoD GTPases regulate receptor activity in Semaphorin 3A-induced cytoskeletal collapse. J. Neurosci. Off. J. Soc. Neurosci. *22*, 471–477.

Atypical Rho GTPases in health and disease

2

Pontus Aspenström*

Department of Microbiology, Tumor and Cell Biology, Karolinska Institutet,
Nobels väg 16, Box 280, Stockholm SE-171 77, Sweden

*pontus.aspenstrom@ki.se

Keywords: Atypical Rho GTPases, RhoD, RhoU, actin, stress fiber,
cell migration.

2.1. Introduction

This chapter describes the concept of atypical Rho GTPases and their roles in
normal physiology as well as in conditions of disease. Before we touch upon the
subject of atypical Rho GTPases, we need to contemplate on what typical
Rho GTPases. It might seem a philosophical question but in the case of Rho
GTPases, the typical Rho GTPases (they can also be referred to as classical Rho
GTPases), they behave similarly to the founding fathers of the small GTPase
superfamily, the Ras GTPases. These small enzymes are known to bind and hydrolyze GTP, and in the process, they change their three-dimensional fold from an
"active" GTP-bound conformation to an "inactive" GDP-bound conformation
(Jaffe and Hall, 2005). Activation is achieved by the exchange of the hydrolyzed
GDP to a new fresh GTP, leading to the activated conformation being restored and
the completion of the GTPase cycle. This cycling between active and inactive
conformation is a rather slow process, but it can be speeded up by guanine nucleotide exchange factors (GEFs) that can catalyze the exchange of GDP for GTP and

by GTPase activating proteins (GAPs), which act as auxiliary factors and speed up the low intrinsic GTPase activity manifold (Hodge and Ridley, 2016). This way, the GEFs act as positive regulators and the GAPs as negative regulators of Rho GTPases (see Chapter 3 by Amin and Ahmadian). This scheme is true for the Ras GTPases and for the typical Rho GTPases. In addition, the Rho GTPases differ from Ras by having a third category of regulatory proteins, the GDP dissociation inhibitors (GDIs) that can sequester the typical Rho GTPases in the inactive conformation (Garcia-Mata *et al.*, 2011).

Now, what about the atypical Rho GTPases, what is making them so special? The simple answer is that they do not follow the same GDP/GTP cycling pattern of their sibling typical Rho GTPases (Aspenström *et al.*, 2007). There are basically two different mechanisms underlying their atypicality, but both mechanisms impinge on the same basic principle; the atypical Rho GTPase is constitutively GTP-bound and thereby always in the active conformation. The first category of atypical Rho GTPases, we can call them GTPase deficient, has a stalled GTP hydrolysis, or they hydrolyze GTP extremely inefficiently. In contrast, the second category has an intact GTPase activity but also a significantly elevated intrinsic GDP/GTP exchange activity. This means that, once the nucleotide is hydrolyzed, it is immediately replaced by a new GTP. The reason why GDP can be replaced with GTP is because there is 10-fold excess of GTP over GDP in most cells. In summary, both categories of atypical Rho GTPases are likely to reside in a constitutively active conformation at any given time point. This is what makes them atypical. The terminology typical versus atypical should not be confused by normal versus abnormal Rho GTPases. In fact, there are 10 typical and 10 atypical Rho GTPases (Fig. 1(a)), so what is normal?

In this chapter, the focus will be on the second category, the Rho GTPases with high intrinsic GDP/GTP exchange activity. They are sometimes referred to as fast-cycling, but it is probably more appropriate to call them "self-cycling". I will describe the different members of this type of atypical Rho GTPases in more detail, describe potential regulatory mechanisms, and finish by some insights into the involvement of atypical Rho GTPases in human pathologies. The GTPase-deficient atypical Rho GTP have been reviewed in great detail and will only be mentioned in brief terms in the following section (Chardin, 2006).

2.2. GTPase-deficient atypical Rho GTPases

The first protein of this type was described by the group of Jeff Settleman and called RhoE, to follow the alphabet after the by-then-known Rho GTPases RhoA, B, C and D. RhoE was isolated in a yeast two-hybrid screen for proteins interacting

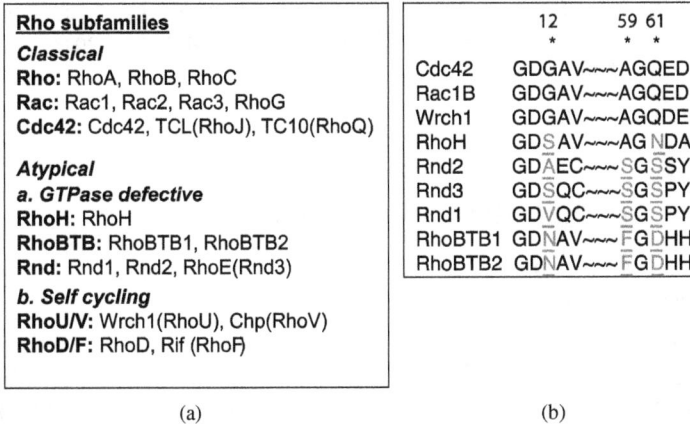

Rho subfamilies
Classical
Rho: RhoA, RhoB, RhoC
Rac: Rac1, Rac2, Rac3, RhoG
Cdc42: Cdc42, TCL(RhoJ), TC10(RhoQ)
Atypical
a. GTPase defective
RhoH: RhoH
RhoBTB: RhoBTB1, RhoBTB2
Rnd: Rnd1, Rnd2, RhoE(Rnd3)
b. Self cycling
RhoU/V: Wrch1(RhoU), Chp(RhoV)
RhoD/F: RhoD, Rif (RhoF)

	12	59	61
	*	*	*
Cdc42	GDGAV~~~AGQED
Rac1B	GDGAV~~~AGQED
Wrch1	GDGAV~~~AGQDE
RhoH	GDSAV~~~AGNDA
Rnd2	GDAEC~~~SGSSY
Rnd3	GDSQC~~~SGSPY
Rnd1	GDVQC~~~SGSPY
RhoBTB1	GDNAV~~~FGDHH
RhoBTB2	GDNAV~~~FGDHH

(a)   (b)

Figure 1. (a) The atypical Rho GTPases. (b) Amino acid sequence over codons 12, 59, and 61 of atypical Rho GTPases.

with p190RhoGAP (Foster *et al.*, 1996). RhoE turned out to have amino acid residues in positions equivalent to 12, 59, and 61 of Ras (Fig. 1(b)). Ras proteins harboring mutations in these positions are GTPase deficient and thus constitutively GTP bound and confer oncogenicity. Therefore, it was not surprising that RhoE was found to possess a very low GTPase activity, if any. This initial observation was followed by one from the groups of Pierre Chardin and Alan Hall, in which two new RhoE-like proteins were characterized (Nobes *et al.*, 1998). This response causes cells to round up; hence the proteins were named Rnd1 and Rnd2.

These proteins are also GTPase defective and were shown to induce stress fiber dissolution and break-down of focal adhesion when microinjected and expressed in Swiss 3T3 fibroblasts. The authors also gave RhoE the alternative name Rnd3 and suggested that the Rnd proteins act as Rho antagonists (Nobes *et al.*, 1998). These observations were followed by the identification of RhoH and the RhoBTBs. RhoH, also known as translocation three four (TTF), was first isolated as a fusion partner with the BCL-6 gene, which is frequently disrupted in B-diffuse large-cell non-Hodgkin's lymphoma (Dallery *et al.*, 1995). RhoH has turned out to be a regulator of hematopoietic cells and lack of RhoH results in T cell deficiency and impaired T cell function (Troeger and Williams, 2013; Crequer *et al.*, 2012). In contrast, the RhoBTB proteins are quite divergent from all other Rho GTPases (see Chapter 1 by Fort). They have N-terminal GTPase-like domains, but in addition, they possess two BTB domains which confer protein–protein interaction. Functionally they are also quite divergent from all Rho GTPases. They have no clear effects on cytoskeletal organization; instead they are subunits in an E3 ubiquitin ligase complex together with Cullin-3 and the R-finger domain-containing protein Rbx1/ROC1 (Ji and Rivero, 2016). The substrates of the RhoBTBs are not known, but they have been suggested to function as tumor suppressors.

2.3. Self-cycling atypical Rho GTPases

The concept of a type of GTPase with an increased intrinsic GDP/GTP exchange activity has emerged from the characterization of a point mutation in Cdc42 encompassing amino acid residue 28 (F28L). This mutation rendered the protein to undergo spontaneous GTP/GDP exchange, and in cultured fibroblasts, this increased "self-cycling" ability was associated with cell transformation seen as reduced contact inhibition and anchorage-independent growth (Lin *et al.*, 1997). The intrinsic exchange activity results in that the protein is predominantly in the active, GTP-bound conformation, because there is an excess of GTP over GDP inside cells. The identification of Wrch-1 (now called RhoU) demonstrated that some Rho GTPases possess an elevated GDP/GTP exchange activity in their wild-type versions (Tao *et al.*, 2001; Saras *et al.*, 2004). Another example of a protein with an elevated intrinsic exchange activity is the cancer-related Rac1 splice-variant Rac1b (Fiegen *et al.*, 2004). Finally, RhoD and Rif, although first identified as classical Rho GTPases, also have an elevated intrinsic GDP/GTP exchange activity (Jaiswal *et al.*, 2013).

2.3.1. *Wrch-1/RhoU and Chp/RhoV*

Wrch-1 was identified as a gene that upregulated in response to Wnt-1, hence the name: Wnt-1-responsive Cdc42 homologue. It was found to be similar to Cdc42 and another Cdc42-like protein called Chp (Cdc42 homologous protein or RhoV) (Aronheim *et al.*, 1998). Wrch-1 and Chp differ from Cdc42 and other Rho GTPases in other ways than by having self-cycling abilities. Both proteins have N-terminal extensions of a type not found in any other members of the Rho GTPases; Wrch-1 has a 46-amino-acid-residue proline-rich extension, and Chp has an extension of 28 amino acid residues (Aronheim *et al.*, 1998; Tao *et al.*, 2001). The proline-rich motif in Wrch-1 has been found to bind SH3 domain-containing proteins such as Nck, Grb2, and PLCγ (Saras *et al.*, 2004; Shutes *et al.*, 2004). Possibly, the extension has a regulatory role and could constitute a way to control the activity of Wrch-1.

As mentioned previously, Wrch-1 lacks regulatory RhoGEFs and RhoGAPs, which raises questions regarding its regulation. It is clear that Wrch-1 is under transcriptional control, and its transcription can, for instance, be triggered by Wnt-1, RANKL, as well as by gp130-binding cytokines via a STAT3-dependent mechanism (Tao *et al.*, 2001; Brazier *et al.*, 2006; Schiavone *et al.*, 2009). There are several observations suggesting that the N-terminal proline-rich extension of Wrch-1 can have a regulatory role. It is clear that there is no difference in GTP hydrolysis and GDP/GTP exchange activity between the full length and an N-terminal-truncated Wrch-1, *in vitro*. *In vivo*, however, the N-terminal-truncated mutant has a stronger affinity for the effector protein Pak1, and cells expressing this mutant have an increased ability of anchorage-independent growth (Shutes *et al.*, 2004). SH3 domain-containing proteins, such as Grb2, binding to the proline-rich domain, are likely to have a role in this regulation. Another role of Grb2 is apparently to couple Wrch-1 to the activated EGF receptor and the function of Wrch-1 might be to facilitate the localization of EGFR to endosomes (Zhang *et al.*, 2011) (Fig. 2).

Wrch-1 has a CCFV tetrapeptide at its very C-terminal end. This motif resembles the CAAX motifs found in most Rho GTPases; however, it does not appear to be functional since it does not seem to be modified by isoprenylation.

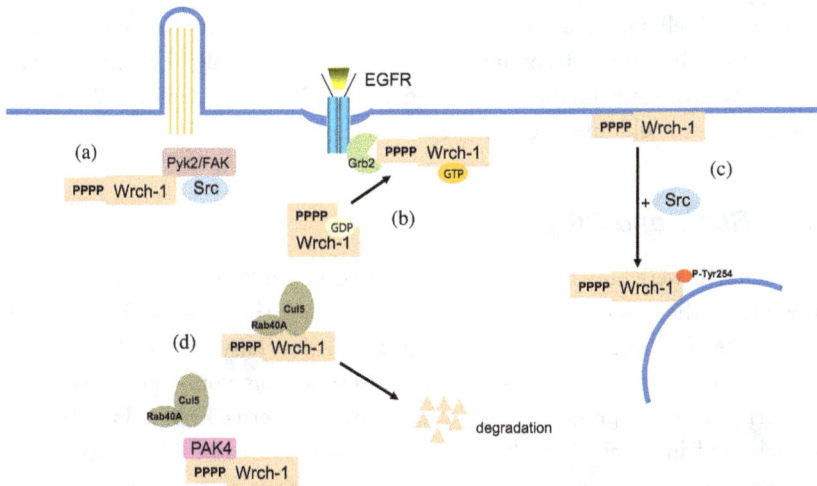

Figure 2. Schematic representation of Wrch-1 function. (a) Wrch-1-triggered filopodia require Pyk2/FAK and Src. (b) SH3 domain-containing proteins, such as Grb2 have roles in Wrch-1 activation. (c) Src-dependent phosphorylation of Wrch-1 at tyrosine residue 254 result in relocalization of Wrch-1 from the plasma membrane to cytoplasmic vesicles. (d) An E3 ubiquitin ligase consisting of Rab40A and cullin5 ubiquitinates Wrch-1 and targets it for degradation. PAK4 binding inhibits the Rab40A/Cul5 complex and stabilizes Wrch-1.

Instead, the antepenultimate cysteine undergoes palmitoylation, and this modification is needed for a correct targeting of Wrch-1 to the plasma membrane. Mutation of this cysteine or treatment of cells with the palmitoylation inhibitor 2-bromohexadecanoic acid results in a relocalization of Wrch-1 to cytosolic vesicles, most likely to endosomes (Berzat *et al.*, 2005). Wrch-1 also appears to be under a negative control by tyrosine phosphorylation. The nonphosphorylated Wrch-1 is localized to the plasma membrane in a GTP-bound conformation and can bind its downstream target protein Pak1. However, Src-dependent phosphorylation of tyrosine residue 254 results in relocalization away from the plasma membrane and the interaction to Pak1 is lost (Alan *et al.*, 2010). Chp also has an N-terminal proline-rich extension, but if it in any way confers a regulatory function is not known.

Ectopic expression of Wrch-1 in fibroblasts, or fibroblast-like cells, results in the formation of thin filopodia. These filopodia are formed in an Src- and Pyk2/FAK-dependent manner (Ruusala and Aspenström, 2008). In one study, Wrch-1 was found to localize to focal adhesions, and depletion of Wrch-1 resulted in increased focal adhesion size, whereas increased Wrch-1 expression resulted in focal-adhesion disassembly (Chuang *et al.*, 2007). A potential explanation for these phenotypes comes from studies on PAK4 and Wrch-1. PAK4 is involved in the regulation of cell adhesion via Wrch-1. This regulation occurs through a mechanism that involves Wrch-1 ubiquitination by the Rab40A-cullin5 complex. PAK4 protects Wrch1 from ubiquitination and degradation since, in the absence of PAK4, Wrch-1 is ubiquitinated and degraded (Dart *et al.*, 2015).

2.3.2. *RhoD and Rif*

RhoD and Rif were both originally classified as classical Rho GTPases; however, a closer look at their kinetic properties revealed that they, similar to Wrch-1 and Chp, have elevated intrinsic GDP/GTP exchange activities (Jaiswal *et al.*, 2013). RhoD is expressed in most tissues and also in several commonly used cell lines. Furthermore, it is expressed at high levels in mouse uterus, liver, kidney, bladder, stomach, and intestine (Murphy *et al.*, 1996; Boureux *et al.*, 2007). Rif is also widely expressed in human tissues, in particular in the colon, stomach, and spleen (Ellis and Mellor, 2000). Rif is better expressed in normal human B cells, compared to its expression in other lymphocytes, which could indicate that Rif has a role in the normal functioning of B cells (Gouw *et al.*, 2005).

Both RhoD and Rif are known to trigger the formation of peripheral protrusions in several cell types (Murphy *et al.*, 1996; Ellis and Mellor, 2000; Aspenström *et al.*, 2004; Blom *et al.*, 2017). The mechanisms for the RhoD- and Rif-induced filopodia are not entirely known, but they do not require the involvement of Cdc42 (Ellis and Mellor, 2000). In the case of Rif, filopodia formation has been suggested to require diaphanous-related formins such as mDia (Pellegrin and Mellor, 2005; Goh *et al*., 2011; Koizumi *et al*., 2012). In neuronal cells, filopodia-like precursors of dendritic spines elongate through a mechanism that requires Rif and mDia2 (Hotulainen *et al*., 2009). Rif was also shown to bind to I-BAR proteins such as IRTKS. I-BAR domains are known to bind lipid bilayers and induce a curvature, for instance, during the initial phase of endocytosis. Proteins with I-BAR domains catalyze the formation of protrusions. In this study, Rif was shown to induce filopodia via a mechanism that, in addition to IRTKS, also involved Eps8 and the WASP family protein WAVE2 (Sudhaharan *et al*., 2016). Thus, it seems as if Rif can use different mechanisms to trigger filopodia, and moreover, it is worth remembering that all filopodia are not equal. Cdc42, Wrch-1, Chp, RhoD, and Rif all can trigger the formation of actin-containing protrusions by slightly different mechanisms (Aspenström *et al*., 2004).

The formation of RhoD-induced filopodia is accompanied by stress fiber dissolution and the appearance of short bundles of actin filaments in several cell types (Gad *et al*., 2012; Blom *et al*., 2017). The formation of these bundles of actin filaments is probably a reflection of the dispersion of focal adhesions that is caused by the expression of constitutively active RhoD^{G26V} (Tsubakimoto *et al*., 1999; Gad *et al*., 2012; Blom *et al*., 2017).

There are several potential mechanisms behind the effect on RhoD on focal-adhesion dynamics (Fig. 3). One plausible mediator is the Ser/Thr protein Zipper kinase (ZIPK, also known as death-associated protein kinase 3), which can regulate focal adhesion kinase (FAK) activity and cell adhesion. ZIPK is a RhoD effector, and constitutively active RhoD^{G26V} can modulate this activity (Nehru *et al*., 2013a, b). Rif appears to have a different effect on stress fiber organization compared to RhoD, since the expression of constitutively active Rif mutant results in stress fiber formation, at least in epithelial cells. These stress fibers require the concerted actions of mDia1 and the Ser/Thr protein kinase Rho kinase (ROCK). However, Rif expression does not result in increased activity of ROCK. Instead, Rif appears to regulate the subcellular localization of ROCK and thereby compartmentalize the signal that regulates myosin activation and the subsequent stress fiber formation (Fan *et al*., 2010).

Since both RhoD and Rif have a strong effect on actin dynamics, it is not surprising to find RhoD to have important roles in regulating cell migration. Transfection of the constitutively active RhoD^{G26V} in fibroblasts resulted in decreased cell migration, measured by the so-called phagokinetic track assay (Tsubakimoto *et al*., 1999). Furthermore, it has been shown that RhoD^{G26V}-

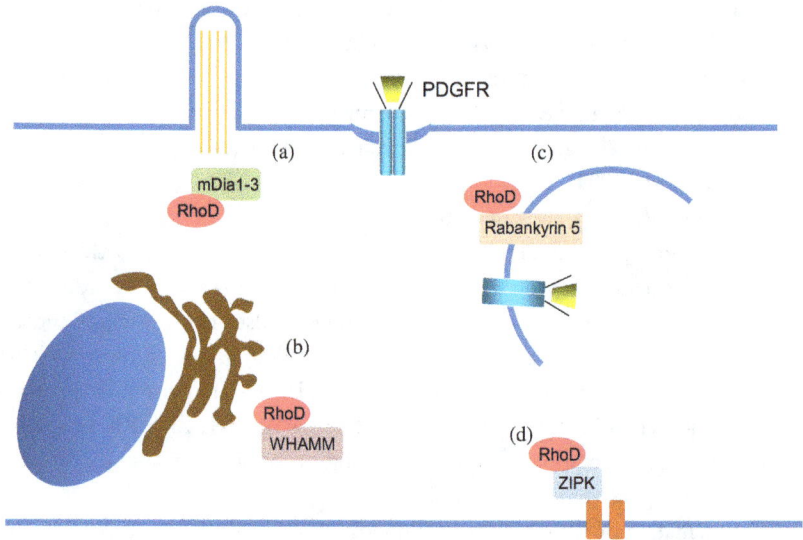

Figure 3. Schematic representation of RhoD function. (a) RhoD-induced filopodia requires diaphanous-related formins. (b) RhoD binds the Golgi component WHAMM and has a role in protein transport from ER-Golgi to the plasma membrane. (c) RhoD and its effector Rabankyrin-5 have roles in the internalization of the PDGFRβ. (d) RhoD affects focal-adhesion dynamics, possibly through its effector ZIPK.

expressing endothelial cells, in essence, are immotile, both in the absence and presence of FGF (Murphy *et al.*, 2001). Expression of the active RhoD mutant in glioblastoma cells also results in less dynamic actin at the cell edges (Blom *et al.*, 2017). The reverse situation, RhoD ablation, had almost the opposite effect: fibroblast lacking RhoD moved slightly longer distance, but the migration was less persistent. In fact, chemotaxis toward a gradient of platelet-derived growth factor (PDGF) was hampered in RhoD depleted cells (Blom *et al.*, 2017). These observations indicate that RhoD is needed for a correct cell polarization and directed cell migration.

2.3.3. *RhoD in the control of endosome motility and receptor internalization*

RhoD was already from its initial characterization implicated in the processes that integrate early endosome motility and actin (Murphy *et al.*, 1996; Ellis and Mellor, 2000; Murphy *et al.*, 2001; Gasman *et al.*, 2003). Transiently transfected RhoD localizes to Rab5-positive cytoplasmic vesicles, possibly corresponding to early endosomes. The enlargement and perinuclear accumulation of endosomes that is induced by overexpression of a constitutively active Rab5^{Q79L} was inhibited by simultaneous expression of RhoD^{G26V} (Murphy *et al.*, 1996). Interestingly, RhoD has been ascribed a role in endosomal trafficking of the Src-family kinases. Knock-down of RhoD was shown to result in loss of the endosome-targeted Fyn, which suggested a role for RhoD in the regulation of Src-family kinases (Sandilands *et al.*, 2007).

A yeast two-hybrid screen identified the Rab5 effector Rabankyrin-5 as a RhoD-binding protein. RhoD and Rabankyrin-5 colocalize to Rab5-positive endosomes, which suggests a role for Rabankyrin-5 in the coordination of RhoD and Rab5 in endosomal trafficking (Nehru *et al.*, 2013a, b). There is a reciprocal relationship between RhoD and Rabankyrin-5 in endosome trafficking, since knock-down of either RhoD or Rabankyrin-5 results in the mislocalization of the other. The biological context in which RhoD and Rabankyrin-5 collaborate appears to relate to trafficking of receptor tyrosine kinases. Knock-down of RhoD can interfere with the internalization of the PDGF-β receptor and the subsequent activation of its downstream signaling cascades, which is one possible reason for the defective chemotaxis of RhoD-depleted cells (Schnatwinkel *et al.*, 2004; Nehru *et al.*, 2013a, b; Blom *et al.*, 2017). This suggests that RhoD and Rabankyrin-5 are likely to have roles in the coordination of RhoD and Rab activities, for instance, during internalization and trafficking of transmembrane receptors.

2.3.4. *RhoD in cytokinesis*

Expression of constitutively active RhoD^{G26V} in several cell types, including Balb/3T3 and C3/10T1/2 fibroblasts and N1E-115 neuroblastoma cells, results in multinucleated cells, which suggests a defect in cytokinesis. Defective cleavage was also seen in *Xenopus* embryos injected with RhoD^{G26V} (Tsubakimoto *et al.*, 1999). Transgenic mice expressing RhoD^{G26V} exclusively at the basal layer of the epidermis show hyperplasia and perturbed differentiation of epidermal cells (Kyrkou *et al.*, 2013). Obviously, RhoD has an important role in S-phase entry, as the knock-down of RhoD in endothelial cells results in the accumulation of cells in G1. Overexpression of RhoD^{G26V} results in increased cell proliferation and also in aberrant centrosome amplification (Kyrkou *et al.*, 2012). Knock-down of RhoD does not interfere with cytokinesis but results in significantly longer cell cycle (Blom *et al.*, 2017). This is another example to illustrate that precise RhoD activation is needed for normal cell physiology; both RhoD overactivity and underactivity result in derailed cell proliferation.

2.4. Activating point mutations in Rac1

In addition to the "natural" self-cycling members of the atypical Rho GTPases, there is an increasing awareness that mutants of the classical Rho GTPases can in fact have elevated intrinsic GDP/GTP cycling activity. One example was the aforementioned $Cdc42^{F28L}$ mutant (Lin *et al.*, 1997). Another example on the same theme is the panel of Rac1 point mutations that have been identified, predominantly in melanoma cancer (Kawazu *et al.*, 2013; Halaban, 2015). Rho GTPases have been considered to be refractive to being mutated in cancer. Instead, RhoGEFs and RhoGAPs have been implicated as targets for cancer-associated mutations (Hodge and Ridley, 2016). But recent findings, the fruit of the increased power of next generation sequencing, have changed this paradigm (see Chapter 7 by Der). Studies on melanoma have shown that the $Rac1^{P29S}$ found in about 4–7% of all melanomas (Krauthammer *et al.*, 2012; Halaban, 2015). It is the third most recurrent mutation in melanoma after the $B-Raf^{V600E}$ and $N-Ras^{Q61K/R}$ mutations. The $Rac1^{P29S}$ has a significantly increased intrinsic GDP/GTP cycling ability, and this is most likely a driver mutation making Rac1 an oncogene (Davis *et al.*, 2013). Another example of a cancer-associated Rac1 variant is the Rac1b splice variant. It is caused by alternative splicing event resulting in 19 extra amino acid residues after the Switch II motif in Rac1 (Jordan *et al.*, 1999; Singh *et al.*, 2004). This splice form is presumably expressed at low level in normal cells, but its expression level is increased in a number of human cancers and the Rac1b splice variant actually can antagonize the activity of the wild-type Rac1 (Jordan *et al.*, 1999; Matos *et al.*, 2003; Mori *et al.*, 2013). It is thus clear that "self-cycling" atypical Rho GTPases or mutations of classical Rho GTPases are associated with tumor promotion, presumably because the proteins are constitutively active in the true sense of the word. Of note, the normal so-called constitutively active mutants of Rho members, such as $Rac1^{G12V}$ and $Rac1^{Q61L}$, are not really constitutively active; rather, they are GTPase defective, which is not the same thing.

2.5. Summary

The signaling networks involving Rho GTPase have increased in complexity, and the studies on the atypical Rho GTPases have significantly broadened the concept of Rho-regulated biological pathways. There are still many issues that need to be resolved in the future. We certainly need to obtain more information of the mechanisms by which the atypical Rho GTPases are activated and inactivated. Moreover, we need to pinpoint their binding partners in order to get a grip on the mechanisms underlying the signaling capacity of the atypical Rho GTPases. We have already seen that disturbances in their function result in oncogenic transformation. It is likely that dysfunctional atypical Rho GTPases result in other disease conditions, which make it important to increase our attention to these proteins and the biological processes that involve these proteins.

References

Alan, J.K., Berzat, A.C., Dewar, B.J., Graves, L.M., and Cox, A.D. (2010). Regulation of the Rho family small GTPase Wrch-1/RhoU by C-terminal tyrosine phosphorylation requires Src. Mol. Cell. Biol. *30*, 4324–4338.

Aronheim, A., Broder, Y.C., Cohen, A., Fritsch, A., Belisle, B., and Abo A. (1998). Chp, a homologue of the GTPase Cdc42Hs, activates the JNK pathway and is implicated in reorganizing the actin cytoskeleton. Curr. Biol. *8*, 1125–1128.

Aspenström, P., Fransson, Å., and Saras, J. (2004). The Rho GTPases have diverse effects on the organization of the actin filament system. Biochem. J. *377*, 327–337.

Aspenström, P., Ruusala, A., and Pacholsky, D. (2007). Taking the Rho GTPases to the next level: the cellular function of the atypical Rho GTPases. Exp. Cell Res. *313*, 3673–3679.

Berzat, A.C., Buss, J.E., Chenette, E.J., Weinbaum, C.A., Shutes, A., Der, C.J., Minden, A., and Cox, A.D. (2005). Transforming activity of the Rho family GTPase, Wrch-1, a Wnt-regulated Cdc42 homolog, is dependent on a novel carboxyl-terminal palmitoylation motif. J. Biol. Chem. *280*, 33055–33065.

Blom, M., Reis, K., Heldin, J., Kreuger, J., and Aspenström, P. (2017). The atypical Rho GTPase RhoD is a regulator of actin cytoskeleton dynamics and directed cell migration. Exp. Cell Res. *352*, 255–264.

Boureux, A., Vignal, E., Faure, S., and Fort, P. (2007). Evolution of the Rho family of Ras-like GTPases in eukaryotes. Mol. Biol. Evol. *24*, 203–216.

Brazier, H., Stephens, S., Ory, S., Fort, P., Morrison, N., and Blangy, A. (2006). Expression profile of RhoGTPases and RhoGEFs during RANKL-stimulated osteoclastogenesis: identification of essential genes in osteoclasts. J. Bone Miner. Res. *21*, 1387–1398.

Chardin P. (2006). Function and regulation of Rnd proteins. Nat. Rev. Mol. Cell Biol. *7*, 54–62.

Crequer, A., Troeger, A., Patin, E., Ma, C.S., Picard, C., Pedergnana, V., Fieschi, C., Lim, A., Abhyankar, A., Gineau, L., Mueller-Fleckenstein, I., Schmidt, M., Taieb, A., Krueger, J., Abel, L., Tangye, S.G., Orth, G., Williams, D.A., Casanova, J.L., and Jouanguy, E. (2012). Human RHOH deficiency causes T cell defects and susceptibility to EV-HPV infections. J. Clin. Invest. *122*, 3239–3247.

Chuang, Y., Valster, A., Coniglio, S.J., Backer, J.M., and Symons, M. (2007). The atypical Rho family GTPase Wrch-1 regulates focal adhesion formation and cell migration. J. Cell Sci. *120*, 1927–1934.

Dallery, E., Galiègue-Zouitina, S., Collyn-d'Hooghe, M., Quief, S., Denis, C., Hildebrand, M.-P., Lantoine, D., Deweindt, C., Tilly, H., Bastard, C., and Kerckaert, J.-P. (1995).

TTF, a gene encoding a novel small G protein, fuses to the lymphoma-associated LAZ3 gene by t(3;4) chromosomal translocation. Oncogene *10*, 2171–2178.

Dart, A.E., Box, G.M., Court, W., Gale, M.E., Brown, J.P., Pinder, S.E., Eccles, S.A., and Wells, C.M. (2015). PAK4 promotes kinase-independent stabilization of RhoU to modulate cell adhesion. J. Cell Biol. *211*, 863–879.

Davis, M.J., Ha, B.H., Holman, E.C., Halaban, R., Schlessinger, J., and Boggon, T.J. (2013). RAC1P29S is a spontaneously activating cancer-associated GTPase. Proc. Natl. Acad. Sci. USA *110*, 912–917.

Ellis, S. and Mellor, H. (2000). The novel Rho-family GTPase Rif regulates coordinated actin-based membrane rearrangements. Curr. Biol. *10*, 1387–1390.

Fan, L., Pellegrin, S., Scott A., and Mellor, H. (2010). The small GTPase Rif is an alternative trigger for the formation of actin stress fibers in epithelial cells. J. Cell Sci. *123*, 1247–1252.

Fiegen, D., Haeusler, L.C., Blumenstein, L., Herbrand, U., Dvorsky, R., Vetter, I.R., and Ahmadian, M.R. (2004). Alternative splicing of Rac1 generates Rac1b, a self-activating GTPase. J. Biol. Chem. *279*, 4743–4749.

Foster, R., Hu, K.Q., Lu, Y., Nolan, K.M., Thissen, J., and Settleman, J. (1996). Identification of a novel human Rho protein with unusual properties: GTPase deficiency and *in vivo* farnesylation. Mol. Cell. Biol. *16*, 2689–2699.

Gad, A.K.B, Nehru, V., Ruusala, A., and Aspenström, P. (2012). RhoD regulates cytoskeletal dynamics via the actin-nucleation-promoting factor WHAMM. Mol. Biol. Cell *23*, 4807–4819.

Gasman, S., Kalaidzidis, Y., and Zerial, M. (2003). RhoD regulates endosome dynamics through diaphanous-related formin and Src kinase. Nat. Cell Biol. *5*, 195–204.

Goh, W.I., Sudhaharan, T., Lim, K.B., Sem, K.P., Lau, C.L., and Ahmed, S. (2011). Rif-mDia1 interaction is involved in filopodium formation independent of Cdc42 and Rac effectors. J. Biol. Chem. *286*, 13681–13694.

Gouw, L.G., Reading, N.S., Jenson, S.D., Lim, M.S., and Elenitoba-Johnson, K.S. (2005). Expression of the Rho-family GTPase gene RHOF in lymphocyte subsets and malignant lymphomas. Br. J. Haematol. *129*, 531–533.

Garcia-Mata, R., Boulter, E., and Burridge, K. (2011). The 'invisible hand': regulation of RHO GTPases by RHO-GDIs. Nat. Rev. Mol. Cell Biol. *12*, 493–504.

Halaban, R. (2015). RAC1 and melanoma. Clin. Ther. *37*, 682–685.

Hodge, R.G. and Ridley, A.J. (2016). Regulating Rho GTPases and their regulators. Nat. Rev. Mol. Cell Biol. *17*, 496–510.

Hotulainen, P., Llano, O., Smirnov, S., Tanhuanpää, K., Faix, J., Rivera, C., and Lappalainen, P. (2009). Defining mechanisms of actin polymerization and depolymerization during dendritic spine morphogenesis. J. Cell Biol. *185*, 323–339.

Jaffe, A.B. and Hall, A. (2005). Rho GTPases: biochemistry and biology. Annu. Rev. Cell. Dev. Biol. *21*, 247–269.

Jaiswal, M., Fansa, E.K., Dvorsky, R., and Ahmadian, M.R. (2013). New insight into the molecular switch mechanism of human Rho family proteins: shifting a paradigm. Biol. Chem. *394*, 89–95.

Ji, W. and Rivero, F. (2016). Atypical Rho GTPases of the RhoBTB subfamily: roles in vesicle trafficking and tumorigenesis. Cells 5(2), 28.

Kawazu, M., Ueno, T., Kontani, K., Ogita, Y., Ando, M., Fukumura, K., Yamato, A., Soda, M., Takeuchi, K., Miki, Y., Yamaguchi, H., Yasuda, T., Naoe, T., Yamashita, Y., Katada, T., Choi, Y.L., and Mano, H. (2013). Transforming mutations of RAC guanosine triphosphatases in human cancers. Proc. Natl. Acad. Sci. USA 110, 3029–3034.

Jordan, P., Brazåo, R., Boavida, M.G., Gespach, C., and Chastre, E. (1999). Cloning of a novel human Rac1b splice variant with increased expression in colorectal tumors. Oncogene 18, 6835–6839.

Koizumi, K., Takano, K., Kaneyasu, A., Watanabe-Takano, H., Tokuda, E., Abe, T., Watanabe, N., Takenawa, T., and Endo, T. (2012). RhoD activated by fibroblast growth factor induces cytoneme-like cellular protrusions through mDia3C. Mol. Biol. Cell 23, 4647–4661.

Krauthammer, M., Kong, Y., Ha, B.H., Evans, P., Bacchiocchi, A., McCusker, J.P., Cheng, E., Davis, M.J., Goh, G., Choi, M., Ariyan, S., Narayan, D., Dutton-Regester, K., Capatana, A., Holman, E.C., Bosenberg, M., Sznol, M., Kluger, H.M., Brash, D.E., Stern, D.F., Materin, M.A., Lo, R.S., Mane, S., Ma, S., Kidd, K.K., Hayward, N.K., Lifton, R.P., Schlessinger, J., Boggon, T.J., and Halaban, R. (2012). Exome sequencing identifies recurrent somatic RAC1 mutations in melanoma. Nat. Genet. 44, 1006–1014.

Kyrkou, A., Soufi, M., Bahtz, R., Ferguson, C., Bai, M., Parton, R.G., Hoffmann, I., Zerial, M., Fotsis, T., and Murphy, C. (2013). RhoD participates in the regulation of cell-cycle progression and centrosome duplication. Oncogene 32, 1831–1842.

Lin, R., Bagrodia. S., Cerione, R., and Manor, D. (1997). A novel Cdc42Hs mutant induces cellular transformation. Curr. Biol. 7, 794–797.

Matos, P., Collard, J.G., and Jordan, P. (2003). Tumor-related alternatively spliced Rac1b is not regulated by Rho-GDP dissociation inhibitors and exhibits selective downstream signaling. J. Biol. Chem. 278, 50442–50448.

Mori, Y., Yagi, S., Sakurai, A., Matsuda, M., and Kiyokawa, E. (2013). Insufficient ability of Rac1b to perturb cystogenesis. Small GTPases 4, 9–15.

Murphy, C., Saffrich, R., Grummt, M., Gournier, H., Rybin, V., Rubino, M., Auvinen, P., Lütcke, A., Parton, R.G., and Zerial, M. (1996). Endosome dynamics regulated by a Rho protein. Nature 384, 427–432.

Murphy, C., Saffrich, R., Olivo-Marin, J.C., Giner, A., Ansorge, W., Fotsis, T., and Zerial, M. (2001). Dual function of RhoD in vesicular movement and cell motility. Eur. J. Cell Biol. 80, 391–398.

Nehru, V., Almeida, F.N., and Aspenström, P. (2013a). Interaction of RhoD and ZIP kinase modulates actin filament assembly and focal adhesion dynamics. Biochem. Biophys. Res. Commun. 433, 163–169.

Nehru, V., Voytyuk, O., Lennartsson, J., and Aspenström, P. (2013b). RhoD binds the Rab5 effector Rabankyrin-5 and has a role in trafficking of the platelet-derived growth factor receptor. Traffic 14, 1242–1254.

Nobes, C.D., Lauritzen, I., Mattei, M.G., Paris, S., Hall, A., and Chardin, P. (1998). A new member of the Rho family, Rnd1, promotes disassembly of actin filament structures and loss of cell adhesion. J. Cell Biol. *141*, 187–197.

Pellegrin, S. and Mellor, H. (2005). The Rho family GTPase Rif induces filopodia through mDia2. Curr. Biol. *15*, 129–133.

Ruusala, A. and Aspenström, P. (2008). The atypical Rho GTPase Wrch1 collaborates with the nonreceptor tyrosine kinases Pyk2 and Src in regulating cytoskeletal dynamics. Mol. Cell. Biol. *28*, 1802–1814.

Saras, J., Wollberg, P., and Aspenström, P. (2004). Wrch1 is a GTPase-deficient Cdc42-like protein with unusual binding characteristics and cellular effects. Exp. Cell Res. *299*, 356–369.

Sandilands, E., Brunton, V.G., and Frame, M.C. (2007). The membrane targeting and spatial activation of Src, Yes and Fyn is influenced by palmitoylation and distinct RhoB/RhoD endosome requirements. J. Cell Sci. *120*, 2555–2564.

Schiavone, D., Dewilde, S., Vallania, F., Turkson, J., Di Cunto, F., and Poli, V. (2009). The RhoU/Wrch1 Rho GTPase gene is a common transcriptional target of both the gp130/STAT3 and Wnt-1 pathways. Biochem. J. *421*, 283–292.

Schnatwinkel, C., Christoforidis, S., Lindsay, M.R., Uttenweiler-Joseph, S., Wilm, M., Parton, R.G., and Zerial, M. (2004). The Rab5 effector Rabankyrin-5 regulates and coordinates different endocytic mechanisms. PLoS Biol. *2*, E261.

Shutes, A., Berzat, A.C., Cox, A.D., and Der, C.J. (2004). Atypical mechanism of regulation of the Wrch-1 Rho family small GTPase. Curr. Biol. *14*, 2052–2056.

Singh, A., Karnoub, A.E., Palmby, T.R., Lengyel, E., Sondek, J., and Der, C.J. (2004). Rac1b, a tumor associated, constitutively active Rac1 splice variant, promotes cellular transformation. Oncogene *23*, 9369–9380.

Sudhaharan, T., Sem, K.P., Liew, H.F., Yu, Y.H., Goh, W.I., Chou, A.M., and Ahmed, S. (2016). The Rho GTPase Rif signals through IRTKS, Eps8 and WAVE2 to generate dorsal membrane ruffles and filopodia. J. Cell Sci. *129*, 2829–2840.

Tao, W., Pennica, D., Xu, L., Kalejttaoa, R.F., and Levine, A.J. (2001). Wrch-1, a novel member of the Rho gene family that is regulated by Wnt-1. Genes Dev. *15*, 1796–1807.

Troeger, A. and Williams, D.A. (2013). Hematopoietic-specific Rho GTPases Rac2 and RhoH and human blood disorders. Exp. Cell Res. *319*, 2375–2383.

Tsubakimoto, K., Matsumoto, K., Abe, H., Ishii, J., Amano, M., Kaibuchi, K., and Endo, T. (1999). Small GTPase RhoD suppresses cell migration and cytokinesis. Oncogene *18*, 2431–2440.

Zhang, J.S., Koenig, A., Young, C., and Billadeau, D.D. (2011). GRB2 couples RhoU to epidermal growth factor receptor signaling and cell migration. Mol. Biol. Cell. *22*, 2119–2130.

Regulators of Rho signaling   **3**

Ehsan Amin* and Mohammad R. Ahmadian[†,‡]

**Institute of Neural and Sensory Physiology, Medical Faculty
of the Heinrich-Heine University, Düsseldorf, Germany*

*[†]Institute of Biochemistry and Molecular Biology II, Medical Faculty
of the Heinrich-Heine University, Düsseldorf, Germany*

[‡]reza.ahmadian@uni-duesseldorf.de

Keywords: Arginine finger, catalytic domain, Dbl family, G domain, membrane association, molecular switch, multidomain proteins, nucleotide binding, RhoGAP, RhoGDI, RhoGEF

3.1. General introduction

The Rho (Ras homolog) family is an integral part of the Ras superfamily of guanine nucleotide-binding proteins (GNBPs). Rho family proteins are crucial for several reasons: (i) Approximately 1% of human genome encodes proteins that either regulate or are regulated by direct interaction with Rho proteins; (ii) they control almost all fundamental cellular processes in eukaryotes, including morphogenesis, polarity, movement, cell division, gene expression, and cytoskeleton reorganization (Jaffe and Hall, 2005); (iii) they are associated with a series of human diseases (Ellenbroek and Collard, 2007).

The Rho family proteins function as molecular switches in the cell and cycle between a GDP-bound, inactive state and a GTP-bound, active state (Dvorsky and Ahmadian, 2004). The cellular regulation of this cycle involves guanine nucleotide exchange factors (GEFs), which accelerate the intrinsic GDP/GTP exchange, and GTPase activating proteins (GAPs), which stimulate the intrinsic GTP hydrolysis activity (Cherfils and Zeghouf, 2013). A prerequisite of Rho protein function is their membrane association, which is achieved by posttranslational modification by isoprenyl groups. Therefore, Rho proteins underlie a third control mechanism that directs their membrane targeting to specific subcellular sites. This is achieved by the function of guanine nucleotide dissociation inhibitors (GDIs), which bind selectively to prenylated Rho proteins and control their cycle between cytosol and membrane. Activation of Rho proteins results in their association with effector molecules that subsequently activate a wide variety of downstream signaling cascades, thereby regulating many important physiological and pathophysiological processes in eukaryotic cells (Etienne-Manneville and Hall, 2002).

3.2. Rho family and the molecular switch mechanism

Members of the Rho family have emerged as key regulatory molecules that couple changes in the extracellular environment to intracellular signal transduction pathways. So far, 20 members of the Rho family have been identified in human, which can be divided in distinct subfamilies based on their sequence homology: Rho (RhoA, RhoB, RhoC), Rac (Rac1 Rac2, Rac3, RhoG), Cdc42 (Cdc42, TC10/RhoQ, TCL/RhoJ, RhoU/Wrch1, RhoV/Chp), RhoD (RhoD, Rif/RhoF), Rnd (Rnd1, Rnd2, Rnd3), RhoH/TTF, RHOBTB (RHOBTB1, RHOBTB2) (Jaiswal *et al.*, 2013; Boureux *et al.*, 2007) (see Chapters 1 and 2 by Fort and Aspenström).

Rho family proteins are approximately 21–25 kDa in size. They typically contain a conserved GDP/GTP binding domain, called G domain, and a C-terminal hypervariable region ending with a consensus sequence known as CAAX (C is cysteine, A is any aliphatic amino acid, and X is any amino acid). The G domain consists of five conserved sequence motifs (G1–G5) (Fig. 1a) that are involved in nucleotide binding and hydrolysis (Bourne *et al.*, 1990; Wittinghofer and Vetter, 2011). In the cycle between the inactive and active states, at least two regions of the protein, switch I (G2) and switch II (G3), undergo structural rearrangements and transmit the "OFF" to "ON" signal (Fig. 1) (Dvorsky and Ahmadian, 2004).

Subcellular localization, which is known to be critical for biological activity of Rho proteins, is achieved by a series of posttranslational modifications at a cysteine residue in the CAAX motif, including isoprenylation (geranylgeranyl or farnesyl), endoproteolysis, and carboxyl methylation (Williams, 2003; Roberts *et al.*, 2008).

A characteristic region of Rho family GTPases is the insert helix (amino acids 124–136, RhoA numbering) that may play a role in effector activation and downstream process (Rose *et al.*, 2005).

Although the majority of the Rho family proteins remarkably are inefficient GTP hydrolyzing enzymes, in quiescent cells, they accumulate in an inactive state

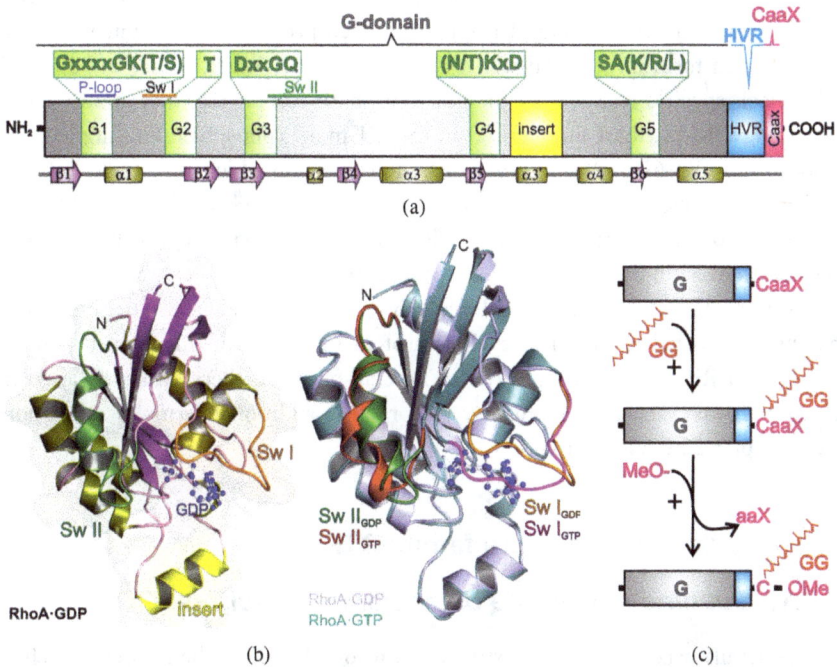

(a)

(b)   (c)

Figure 1. Structural features of Rho family. (a) Primary structure. Schematic drawing of a typical Rho protein with G domain (gray), Rho insert (yellow), C-terminal hypervariable region (HVR: blue) and the Caax-motif (pink). The five conserved G-box motifs (G1–G5 in green) for guanine nucleotide binding are shown above with their consensus sequence. Secondary structural elements (α helices: olive cylinders, β strands: purple arrows) have been represented at the bottom. (b) Tertiary structure of G domain. The three-dimensional structure of Rho proteins represented as cartoon on RhoA structure (PDB code: 1FTN) (left panel). The structural alignment of Rho·GDP (PDB code: 1FTN) and Rho·GTP (PDB code: 1A2B), represented as cartoon (right panel), shows the shift of switch regions upon nucleotide binding. (c) Posttranslational modifications of Rho family. Letters for amino acids: A, Ala; C, Cys; D, Asp; E, Glu; G, Gly; K, Lys; N, Asn; S, Ser; x, any amino acid.

because the GTP hydrolysis is in average two orders of magnitude faster than the GDP/GTP exchange (Jaiswal et al., 2013). Such different intrinsic activities provide the basis for a two-state molecular switch mechanism, which highly depends on the regulatory functions of GEFs and GAPs. 10 out of 20 members of the Rho family belong to these classical molecular switches, namely RhoA, RhoB, RhoC, Rac1, Rac2, Rac3, RhoG, Cdc42, TC10, and TCL (Jaiswal et al., 2013).

The atypical Rho family members, including Rnd1, Rnd2, Rnd3, Rac1b, RhoH/TTF, RhoU, RhoD, and RhoF, have been proposed to accumulate in the GTP-bound form in cells (Jaiswal *et al.*, 2013). Rnd1, Rnd2, Rnd3, and RhoH/TTF lack specific features, as they do not share several essential amino acids, including Gly-12 (Rac1 numbering) in the G1 motif (phosphate-binding loop or P loop) and Gln-61 in the G3 motif or switch II region which are critical in GTP hydrolysis. Thus, they can be considered as GTPase-deficient Rho-related GTP-binding proteins (Garavini *et al.*, 2002). RhoD and Rif are involved in the regulation of actin dynamics (Gad and Aspenström, 2010) and exhibit a strikingly faster nucleotide exchange than GTP hydrolysis. Wrch1, a Cdc42-like protein that has been reported to be a fast cycling protein resembles in this context Rac1b, RhoD, and Rif (Jaiswal *et al.*, 2013). These atypical members do not follow the classical switch mechanism and may thus require additional forms of regulation (see Chapter 2 by Aspenström).

3.3. Regulation of the Rho family GTPases

3.3.1. *Guanine nucleotide dissociation inhibitors (GDIs)*

It is generally accepted that in resting cells, RhoGDIs target the isoprenyl anchor and sequester Rho proteins from their site of action at the membrane in the cytosol (Garcia-Mata *et al.*, 2011).

RhoGDIs undergo a high-affinity interaction with the Rho proteins using an N-terminal regulatory arm contacting the switch regions and a C-terminal domain binding the isoprenyl group (Tnimov *et al.*, 2012). In contrast to the large number of RhoGEFs and RhoGAPs, there are only three known RhoGDIs in human. RhoGDI-1 (also called RhoGDIα) is ubiquitously expressed, whereas RhoGDI-2 (RhoGDIβ) is predominantly found in hematopoietic tissues and lymphocytes. RhoGDI-3 (RhoGDIγ) is found in the lungs, brain, and testes (Adra *et al.*, 1997).

Despite intensive research over the last two decades, the molecular basis by which GDI proteins associate and extract the Rho GTPases from the membrane remains to be investigated. The neurotrophin receptor p75 (p75NTR) and ezrin/radixin/moesin (ERM) proteins have been proposed to displace the Rho proteins from the RhoGDI complex resulting in reassociation with the cell membrane (Yamashita and Tohyama, 2003). Notably, RhoGDI has been shown to be phosphorylated by serine/threonine p21-activated kinase 1 (PAK1), protein kinase A (PKA), protein kinase C (PKC), and the tyrosine kinase Src, thereby decreasing the ability of RhoGDI to form a complex with the Rho proteins, including RhoA, Rac1, and Cdc42 (DerMardirossian *et al.*, 2006).

3.3.2. *Guanine nucleotide exchange factors (GEFs)*

GEFs are able to selectively bind to their respective Rho proteins and accelerate the exchange of tightly bound GDP for GTP (Jaiswal *et al.*, 2013). A common mechanism utilized by GEFs is to strongly reduce the affinity of the bound GDP, leading to its displacement and the subsequent association with GTP (Cherfils and Chardin, 1999; Guo *et al.*, 2005). This reaction involves several stages, including an intermediate state of the GEF in the complex with the nucleotide-free Rho protein. This intermediate does not accumulate in the cell and rapidly dissociates because of the high intracellular GTP concentration leading to the formation of the active Rho·GTP complex. The main reason therefore is that the binding affinity of nucleotide-free Rho protein is significantly higher for GTP than for the GEF proteins (Cherfils and Chardin, 1999; Hutchinson and Eccleston, 2000). Cellular activation of the Rho proteins and their cellular signaling can be selectively uncoupled from the GEFs by overexpressing dominant negative mutants of the Rho proteins (e.g., threonine 19 in RhoA to asparagine) (Heasman and Ridley, 2008). Dominant negative mutants form a tight complex with their cognate GEFs and thus prevent them from activating the endogenous Rho proteins.

3.3.2.1. *Dbl family GEFs*

RhoGEFs of the diffuse B-cell lymphoma (Dbl) family directly activate the proteins of the Rho family (Cook *et al.*, 2013). The prototype of this GEF family is the Dbl protein, which was isolated as an oncogenic product from diffuse B-cell lymphoma cells in an oncogene screen (Eva and Aaronson, 1985) and has been later reported to act on Cdc42 (Hart *et al.*, 1991). Human Dbl family proteins have recently been grouped into functionally distinct categories based on both their catalytic efficiencies and their sequence–structure relationship (Jaiswal *et al.*, 2013). The members of the Dbl family are characterized by a unique Dbl homology (DH) domain (except Trio and Kalirin, which have two DH domains) (Aittaleb *et al.*, 2010; Hoffman and Cerione, 2002; Jaiswal *et al.*, 2011; Viaud *et al.*, 2012).

The DH domain is a highly efficient catalytic machine (Rossman *et al.*, 2005) that is able to accelerate the nucleotide exchange of Rho proteins up to 107-fold. The DH domain is often followed by a pleckstrin homology (PH) domain indicating an essential and conserved function. A model for PH domain-assisted nucleotide exchange has been proposed for some GEFs, such as Dbl, Dbs, and Trio (Rossman *et al.*, 2005). Herein the PH domain serves multiple roles in signaling events anchoring GEFs to the membrane (via phosphoinositides) and directing them toward their interacting GTPases which are already localized to the membrane (Rossman *et al.*, 2005).

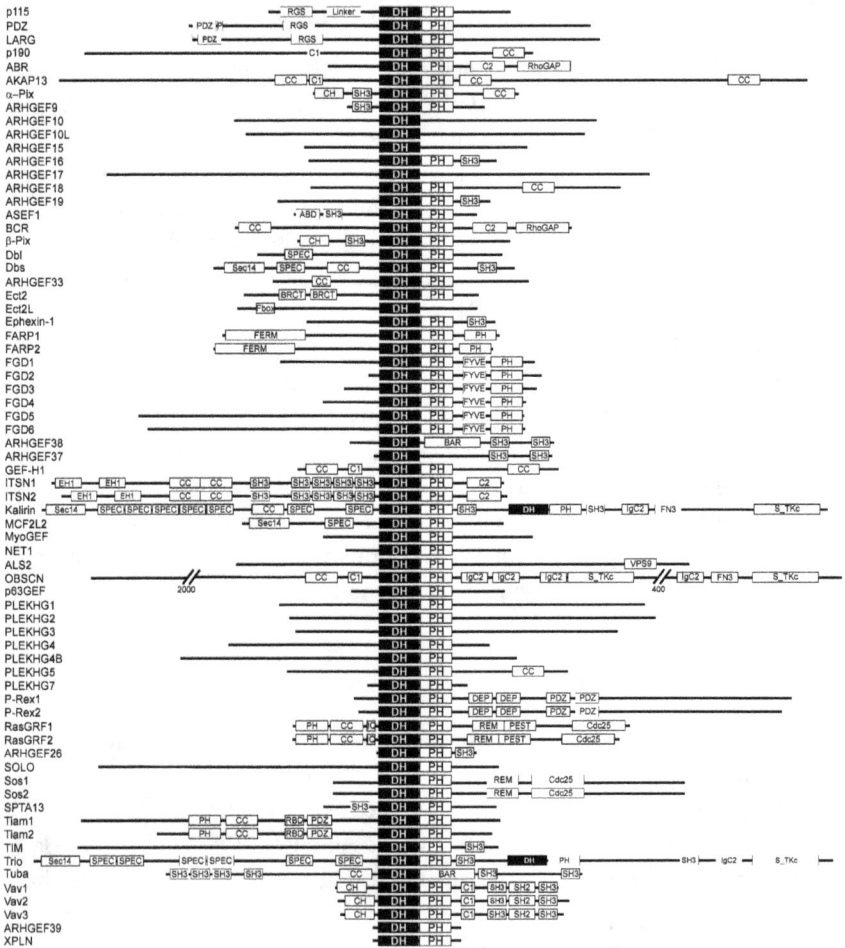

Figure 2. Schematic overview of domain organization of Dbl family proteins. The schematic representation of the domain organization of Dbl family GEFs are illustrated approximately to scale. DH domains almost always occur together with a PH domain at C terminal. Some Dbl proteins contain two DH–PH cassettes while some Dbl proteins lack tandem PH domain. Abbreviations used for various other functional domains.

By searching for DH-domain-containing proteins in the human genome, 70 Dbl proteins have been identified (Jaiswal *et al.*, 2013) (Fig. 2). Interestingly, eight of them lack the C-terminal tandem PH domain, of which three contain a membrane bending and tubulating BAR (Bin/amphiphysin/Rvs) domain, and 7 of 20 investigated proteins did not exhibit any GEF activity (Jaiswal *et al.*, 2013).

In addition to the DH–PH tandem, Dbl family proteins are highly diverse and contain additional domains with different functions, including SH2, SH3, CH, RGS, PDZ, and IQ domains for interaction with other proteins; BAR, PH FYVE, C1, and C2 domains for interaction with membrane lipids; and other functional domains like Ser/Thr kinase, RasGEF, RhoGAP, and RabGEF (Cook *et al.*, 2013). These additional domains have been implicated in autoregulation, subcellular localization, and connection to upstream signals (Dubash *et al.*, 2007; Rossman *et al.*, 2005). Spatiotemporal regulation of the Dbl proteins has been implicated to specifically initiate activation of substrate Rho proteins and to control a broad spectrum of normal and pathological cellular functions (Dubash *et al.*, 2007). Thus, it is evident that members of the Dbl protein family are attractive therapeutic targets for a variety of diseases (Bos *et al.*, 2007; Vigil *et al.*, 2010; see Chapter 9 by Anne Blangy).

3.3.2.2. *Structural and functional characteristic of DH domain*

DH domain is the signature of Dbl family proteins. The catalytic guanine nucleotide exchange activity of Dbl family proteins reside entirely with the DH domain not only sufficient for the catalytic activity but also responsible for the substrate specificity (Jaiswal *et al.*, 2011, 2013). The catalytic DH domain consists of approximately 200 residues and X-ray and NMR analyses of the DH domain of several Dbl proteins reveal that it is composed of unique extended bundle of 10–15 alpha helices (Liu *et al.*, 1998). This helical fold is mainly composed of three conserved regions: CR1, CR2 and CR3, each of them 10–30 residues long form separate alpha helices that pack together (Erickson and Cerione, 2004; Hoffman and Cerione, 2002). The CR1 and CR3 regions are solvent exposed until complexed with Rho proteins (Jaiswal *et al.*, 2013). Beside these three conserved regions (CR1, CR2, and CR3), DH domains of Dbl family share little homology with each other (Jaiswal *et al.*, 2012).

3.3.2.3. *Tandem PH domain of Dbl proteins*

In the majority of Dbl family proteins, the catalytic DH domain is followed by a PH domain of around 100 residues (Fig. 2) and even the identity between PH domains of Dbl family is less than 20%, they share a similar three-dimensional structure with two orthogonal antiparallel β sheets and the C-terminal α helix folds in to cover one end (Lemmon *et al.*, 2002). PH domain was originally identified in number of cytoplasmic signaling proteins that display homology to a region repeated in pleckstrin (Haslam *et al.*, 1993; Tyers *et al.*, 1989). Dbl family comprises DH–PH tandem as signature motif of this family indicating an essential and conserved function for PH domain (Aittaleb *et al.*, 2010; Viaud *et al.*, 2012). The tandem PH domain can act as a "membrane-targeting device" due to its ability to bind phosphoinositides (DiNitto and Lambright, 2006). It can also bind directly to the Rho proteins and potentiate the DH-catalyzed nucleotide exchange reaction (Jaiswal *et al.*, 2011; Liu *et al.*, 1998). In contrast, the PH domains have been shown to bind and to inhibit the activity of the DH domain (Han *et al.*, 1998; Nimnual *et al.*, 1998). Apart from its membrane-targeting properties, emerging evidence suggests that PH domains may also play important regulatory roles by serving as protein–protein interaction modules (Lemmon, 2004).

3.3.2.4. *A plethora of Dbl family proteins*

It has become evident that Dbl family proteins are more abundant and varies in the cells than Rho family proteins. Up to now, 70 Dbl proteins have been reported in human, which are subdivided into different subfamilies: 46 Dbl proteins are mono-specific for Rho-, Rac-, and Cdc42-selective proteins, five are bispecific for Rho- and Cdc42-selective proteins, and six are oligospecific for all three Rho protein subgroups (Jaiswal *et al.*, 2013). Since there are many more Dbl proteins and many of them can activate more than one Rho protein, the activation of Rho proteins catalyzed by Dbl family proteins constitutes a level of regulation in which the signaling pathways can converge or diverge toward one or more Rho proteins (Etienne-Manneville and Hall, 2002). This suggests that at least one representative of each Dbl subfamily is expressed in all mammalian cells and that they may act at distinct subcellular sites.

3.3.3. *GTPase activating proteins (GAPs)*

Hydrolysis of the bound GTP is the timing mechanism that terminates signal transduction of the Rho family proteins and returns them to their inactive, GDP-bound state (Jaiswal *et al.*, 2012). The intrinsic GTPase reaction is usually slow but can be stimulated by several orders of magnitude through interaction with Rho-specific GAPs (Eberth *et al.*, 2005; Fidyk and Cerione, 2002). The RhoGAP family is defined by the presence of a conserved catalytic GAP domain that is sufficient for the interaction with Rho proteins and mediating accelerated catalysis (Amin *et al.*, 2016; Scheffzek and Ahmadian, 2005). The GAP domain supplies a conserved arginine residue, termed "arginine finger", into the GTP-binding site of the cognate Rho protein, in order to stabilize the transition state and catalyze the

GTPase reaction (Amin *et al.*, 2016; Nassar *et al.*, 1998; Rittinger *et al.*, 1997b). This mechanism is utilized by other small GTP-binding proteins, including Ras, Rab, and Arf, although the sequence and folding of the respective GAP families are different (Scheffzek and Ahmadian, 2005; Scheffzek *et al.*, 1997). Masking the catalytic arginine finger is an elegant mechanism for the inhibition of the GAP activity. This has been recently shown for the tumor suppressor protein DLC1, a RhoGAP, which is competitively and selectively inhibited by the SH3 domain of p120RasGAP (Jaiswal *et al.*, 2014; Yang *et al.*, 2009).

The first RhoGAP, p50RhoGAP, was identified by biochemical analysis of human spleen cell extracts in the presence of recombinant RhoA (Garrett *et al.*, 1989). The majority of the RhoGAP family members are frequently accompanied by several other functional domains and motifs implicated in tight regulation and membrane targeting (Amin *et al.*, 2016; Eberth *et al.*, 2009; Moon and Zheng, 2003). Numerous mechanisms have been shown to affect the specificity and the catalytic activity of the RhoGAPs, e.g., intramolecular autoinhibition (Eberth *et al.*, 2009), posttranslational modification (Minoshima *et al.*, 2003), and regulation by interaction with lipid membrane (Ligeti *et al.*, 2004) and proteins (Yang *et al.*, 2009).

RhoGAP insensitivity has been frequently analyzed by the substitution of either amino acids critical for the GTP hydrolysis in Rho proteins, e.g., Gly-14 and Gln-63 in RhoA, which are known as the constitutive active mutants (Ahmadian *et al.*, 1997; Graham *et al.*, 1999), or substitution of the catalytic arginine of the GAP domain to alanine (Fidyk and Cerione, 2002; Graham *et al.*, 1999). The latter approach is in principle very useful under cell-free condition but not really optimal in the cells because an Arg-to-Ala mutant may provide a similar readout as the wild type; it interferes with downstream signaling by competing with the effector(s) for binding to the Rho proteins. RhoGAP mutants at this site are able to persistently bind to and sequester the target Rho protein. This most likely displays a similar readout as the activity of wild-type RhoGAP. Instead of the catalytic arginine, mutating critical "binding determinants", particularly Lys-319 and Arg-323 (p50 numbering), has been recently recommended (Amin *et al.*, 2016). Charge reversal of these residues most likely leads to loss of RhoGAP association with its substrate Rho proteins and consequently the activity of the GAP domain. This is a tool not only for determining the specificity of RhoGAPs but also for investigating GAP domain-independent function(s) of the RhoGAPs.

3.3.3.1. *RhoGAP family proteins*

The GTPase reaction is of great medical significance, since any disruption of this reaction, caused by inhibitory mutations in genes encoding for the GAP proteins, results in a persistent downstream signaling. The first realization that GTPases needs GAPs for their downregulation came from the finding that microinjection of recombinant GTP-bound Ras into living cells result in faster GTP hydrolysis than *in vitro* (Trahey and McCormick, 1987). Characterization of p50RhoGAP later lead to the identification of other RhoGAP-containing proteins: chimaerin and BCR whose amino acid sequences were related to p50RhoGAP with GAP activity (Diekmann *et al.*, 1991). Since then, more than 66 RhoGAP-containing proteins have been identified in humans (Amin *et al.*, 2016; Lancaster *et al.*, 1994). The RhoGAP family is defined by the presence of conserved catalytic GAP domain which is sufficient for the interaction with Rho proteins, mediating accelerated catalysis (Scheffzek and Ahmadian, 2005). Beside their signature RhoGAP domain, most of the RhoGAP family members are frequently accompanied by several other functional domains (Fig. 3). The majority of them can be classified into the following three major groups: (i) lipid- and membrane-binding domains, (ii) peptide- and protein-interacting domains, and (iii) catalytic domains with enzyme activities. Most widespread domains are PH, CC, P, Src homology 3, and BAR/F-BAR. These domains implicate in regulation, membrane targeting, localization, and potential phosphorylation sites and indicate the complexity in the regulation of GTPase activity. Thirteen GAPs lack any additional putative domain but contain highly variable regions at their N and C termini. It is possible that these regions consist of not yet identified motifs, which may contribute to their specific function in the cell.

3.3.3.2. *Structural and functional characteristic of RhoGAP domain*

The GAP domain of RhoGAP family consists of approximately 190 amino acids and share high sequence homology within the family. RhoGAP domain has no similarities at the amino acid level to RasGAP family members, but they resemble with each other in their tertiary structure (Rittinger *et al.*, 1998; Scheffzek *et al.*, 1998). Comparative structural analysis of RhoGAP domain with other GAPs of Ras families suggest that GAP domains of Ras and Rho families are evolutionary related (Rittinger *et al.*, 1998; Scheffzek *et al.*, 1996) and the catalytic domain of RhoGAPs share a core structural fold. RhoGAP domain is made up of seven α helices. The functional characteristic of RhoGAP domain is a pair of conserved basic residues: catalytic arginine (arginine finger) and lysine (Arg-282 and Lys-319 in p50RhoGAP numbering) (Amin *et al.*, 2016; Barrett *et al.*, 1997).

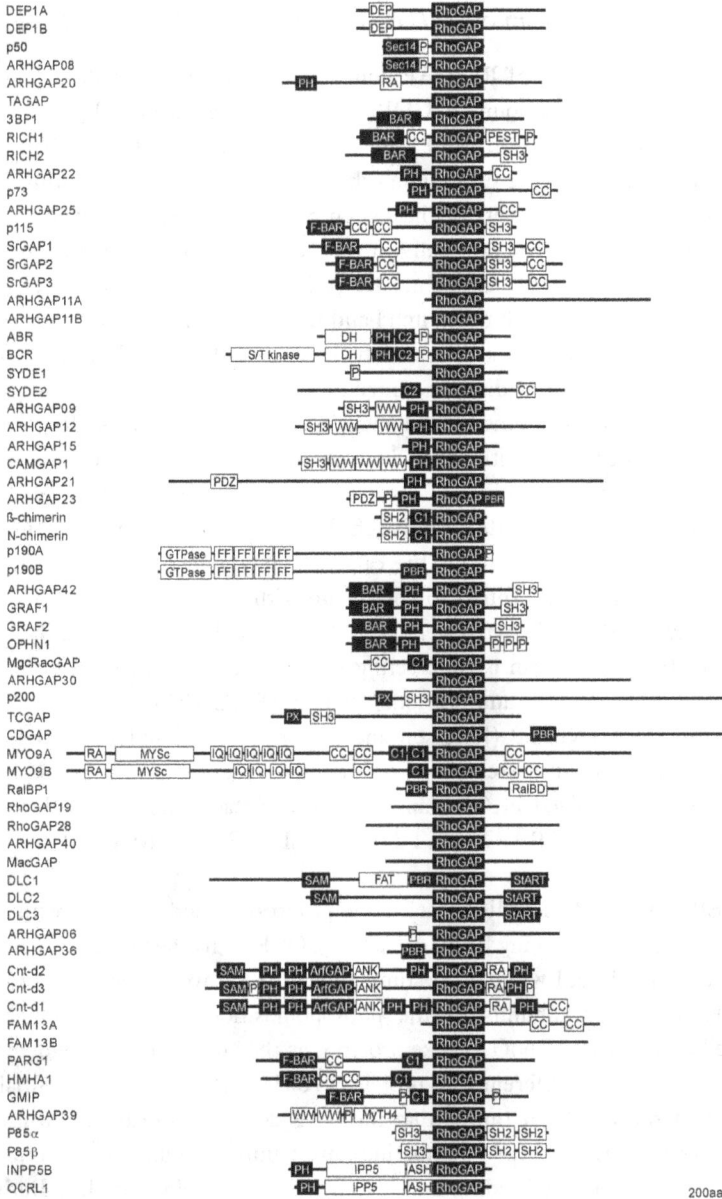

Figure 3. Domain organization of RhoGAPs. The domain organization of RhoGAPs are illustrated approximately to scale. Most of the RhoGAPs are multimodular proteins and have a number of functional domains that may mediate cross talk between Rho proteins and other signaling pathways.

3.3.3.3. *The mechanism of GAP-domain-mediated GTP hydrolysis*

Crystallographic studies of RhoGAP domains in complex with Cdc42 bound to GppNHp, RhoA/Cdc42 bound to GDP·AlF4 (Nassar *et al.*, 1998; Rittinger *et al.*, 1997a; Rittinger *et al.*, 1997b), and RhoA bound to GDP·MgF3 (Graham *et al.*, 2002) gave insight into the catalytic mechanism of GTP hydrolysis stimulation. The GTPase reaction, as part of the switch mechanism, leads to changes in the conformation of GTPases, especially in the flexible and mobile loops known as the switch regions (Dvorsky and Ahmadian, 2004; Vetter and Wittinghofer, 2001). The RhoGAP interacts with the switch I and II regions as well as the Ploop of Rho proteins. The GAP domain accelerates the intrinsic GTP hydrolysis of Rho proteins by two ways. First, it directly contributes to the catalysis by the insertion of the catalytic arginine from the GAP domain into the active site of the Rho protein. This establishes contacts with main-chain carbonyl of Gly-12 (Rac1 numbering) and helps in stabilizing the GTP-hydrolysis transition state (Rittinger *et al.*, 1997a). Second, it stabilizes the negative charges formed during the transition state of GTP hydrolysis and positions the catalytic glutamine residue (Gln-61 Rac1 numbering) of the Rho protein to coordinate with nucleophilic water molecule (Rittinger *et al.*, 1997b; Scheffzek *et al.*, 1998). RhoGAP also stabilizes the switch regions of the Rho protein by interacting with the residues involved in intrinsic GTPases activity (Fidyk and Cerione, 2002). ARHGAP36, CNT-D1, DEP1, DEP2, FAM13B, INPP5P (Jefferson and Majerus, 1995), and OCRL1 lack an arginine finger which makes them catalytically inactive (Amin *et al.*, 2016). ARHGAP36 is involved in Gli transcription factor activation but independent on its GAP domain. ARAP2 (CNT-D1) lacks RhoGAP activity and acts as an ARF6 GAP.

DEP1 and DEP2 coordinate cell cycle progression and interfere with RhoA and signaling despite lacking RhoGAP activity. OCRL1 has been shown to interact with GTP-bound Rac1 without the stimulation of GTP hydrolysis. p85α and p85β (85-kDa regulatory subunits of the phosphoinositide 3-kinases) can also be included on the list of RhoGAP-like proteins, as they do not show any detectable GAP activity toward different Rho proteins (Tolias *et al.*, 1995). An essential prerequisite of the GAP function is that the GAP domain, in order to position its catalytic residue Arg-282 (p50 numbering), must employ a number of amino acids that are responsible for binding and stabilizing the protein complex. Both p85 isoforms lack most of these binding determinants, e.g., Arg-323, Asn-391, Val-394, and Pro-398, along with the conserved amino acids around the arginine finger (Amin *et al.*, 2016).

3.3.3.4. *Overabundance and diversity*

Using database searches, 66 distinct RhoGAP-domain-containing proteins are found to be encoded in human genome, whereas the number of Rho family proteins which need to be regulated by GAPs is 18 (excluding constitutively active Rho proteins). The overabundance of RhoGAPs implies that they must be tightly regulated in the cell in a way that Rho proteins are not accidentally turned off. From 66 RhoGAPs, 57 have a common catalytic domain capable of terminating Rho protein signaling by stimulating the slow intrinsic GTP hydrolysis (GTPase) reaction. Investigation of the sequence–structure–function relationship between RhoGAPs and Rho proteins by combining *in vitro* data with *in silico* data has shown that the RhoGAP domain itself is nonselective and in some cases, rather inefficient under cell-free conditions. This proposes that other domains of RhoGAPs confer substrate specificity and fine-tune their catalytic efficiency in cells.

3.3.3.5. *Regulation and GAP proteins' functions*

RhoGAPs are widely expressed which makes their apparent redundancy questionable. Therefore, it becomes important for the cell to regulate RhoGAPs very tightly to prevent unwanted events that downregulate signaling. To ensure a stringent regulatory control, the RhoGAPs are controlled at different levels indicating that region outside the RhoGAP domain (Fig. 3) must then determine the specificity of RhoGAPs. Numerous mechanisms have been shown to affect the catalytic activity, substrate specificity of RhoGAPs, e.g., autoinhibition (GRAF, OPHN1) (Eberth *et al.*, 2009), posttranslational regulation: phosphorylation (p190GAP, MgcRacGAP) (Minoshima *et al.*, 2003), and regulation by a lipid-binding domain: PH or C2 domain (Ligeti *et al.*, 2004), protein–protein interactions (DLC1/p120RasGAP) (Jaiswal *et al.*, 2014; Yang *et al.*, 2009) and subcellular distribution as specific colocalization of RhoGAPs with Rho proteins at the membrane, e.g., by a scaffolding protein (Bernards, 2003).

3.4. Conclusions

Abnormal activation of Rho proteins has been shown to play a crucial role in cancer, infectious and cognitive disorders, and cardiovascular diseases. However, several tasks have to be yet accomplished in order to understand the complexity of Rho proteins' signaling:

(i) The Rho family comprises of 20 signaling proteins, of which only RhoA, Rac1, and Cdc42 have been comprehensively studied so far. The functions of the other less-characterized members of this protein family await detailed investigation.

(ii) Despite intensive research over the last two decades, the mechanisms by which RhoGDIs associate and extract the Rho proteins from the membrane, and the factors displacing the Rho protein from the complex with RhoGDI remain to be elucidated.

(iii) For the regulation of the 20 Rho proteins, a tremendous number of their regulatory proteins (70 GEFs and 66 GAPs) exists in the human genome. How these regulators selectively recognize their Rho protein targets is not well understood and majority of GEFs and GAPs in humans so far remain uncharacterized.

(iv) Most of the GEFs and GAPs themselves need to be regulated and require activation through the relief of autoinhibitory elements (Eberth *et al.*, 2009; Jaiswal *et al.*, 2011). With a few exceptions (Cherfils and Zeghouf, 2013), it is conceptually still unclear how such autoregulatory mechanisms are operated. A better understanding of the specificity and the mode of action of these regulatory proteins is not only fundamentally important for many aspects of biology but also a master key for the development of drugs against a variety of diseases caused by aberrant functions of Rho proteins.

References

Adra, C.N., Manor, D., Ko, J.L., Zhu, S., Horiuchi, T., Van Aelst, L., Cerione, R.A., and Lim, B. (1997). RhoGDIgamma: a GDP-dissociation inhibitor for Rho proteins with preferential expression in brain and pancreas. Proc. Natl. Acad. Sci. USA *94*, 4279–4284.

Ahmadian, M.R., Mittal, R., Hall, A., and Wittinghofer, A. (1997). Aluminum fluoride associates with the small guanine nucleotide binding proteins. FEBS Letters *408*, 315–318.

Aittaleb, M., Boguth, C.A., and Tesmer, J.J.G. (2010). Structure and function of heterotrimeric G protein-regulated Rho guanine nucleotide exchange factors. Mol. Pharmacol. *77*, 111–125.

Amin, E., Jaiswal, M., Derewenda, U., Reis, K., Nouri, K., Koessmeier, K.T., Aspenström, P., Somlyo, A.V., Dvorsky, R., and Ahmadian, M.R. (2016). Deciphering the molecular and functional basis of RhoGAP family proteins: a systematic approach toward selective inactivation of Rho family proteins. J. Biol. Chem. *291*, 20353–20371.

Barrett, T., Xiao, B., Dodson, E.J., Dodson, G., Ludbrook, S.B., Nurmahomed, K., Gamblin, S.J., Musacchio, A., Smerdon, S.J., and Eccleston, J.F. (1997). The structure of the GTPase-activating domain from p50rhoGAP. Nature *385*, 458–461.

Bernards, A. (2003). GAPs galore! A survey of putative Ras superfamily GTPase activating proteins in man and Drosophila. Biochim. Biophys. Acta *1603*, 47–82.

Bos, J.L., Rehmann, H., and Wittinghofer, A. (2007). GEFs and GAPs: critical elements in the control of small G proteins. Cell *129*, 865–877.

Boureux, A., Vignal, E., Faure, S., and Fort, P. (2007). Evolution of the rho family of ras-like GTPases in eukaryotes. Mol. Biol. Evol. *24*, 203–216.

Bourne, H.R., Sanders, D.A., and McCormick, F. (1990). The GTPase superfamily: a conserved switch for diverse cell functions. Nature *348*, 125–132.

Cherfils, J. and Chardin, P. (1999). GEFs: structural basis for their activation of small GTP-binding proteins. Trends Biochem. Sci. *24*, 306–311.

Cherfils, J. and Zeghouf, M. (2013). Regulation of small GTPases by GEFs, GAPs, and GDIs. Physiol. Rev. *93*, 269–309.

Cook, D.R., Rossman, K.L., and Der, C.J. (2013). Rho guanine nucleotide exchange factors: regulators of Rho GTPase activity in development and disease. Oncogene *16*, 362.

DerMardirossian, C., Rocklin, G., Seo, J.Y., and Bokoch, G.M. (2006). Phosphorylation of RhoGDI by Src regulates Rho GTPase binding and cytosol-membrane cycling. Mol. Biol. Cell *17*, 4760–4768.

Diekmann, D., Brill, S., Garrett, M.D., Totty, N., Hsuan, J., Monfries, C., Hall, C., Lim, L., and Hall, A. (1991). Bcr encodes a GTPase-activating protein for p21rac. Nature *351*, 400–402.

DiNitto, J.P. and Lambright, D.G. (2006). Membrane and juxtamembrane targeting by PH and PTB domains. Biochim. Biophys. Acta *1761*, 850–867.

Dubash, A.D., Wennerberg, K., Garcia-Mata, R., Menold, M.M., Arthur, W.T., and Burridge, K. (2007). A novel role for Lsc/p115 RhoGEF and LARG in regulating RhoA activity downstream of adhesion to fibronectin. J. Cell Sci. *120*, 3989–3998.

Dvorsky, R. and Ahmadian, M.R. (2004). Always look on the bright site of Rho: structural implications for a conserved intermolecular interface. EMBO Rep. *5*, 1130–1136.

Eberth, A., Dvorsky, R., Becker, C.F., Beste, A., Goody, R.S., and Ahmadian, M.R. (2005). Monitoring the real-time kinetics of the hydrolysis reaction of guanine nucleotide-binding proteins. Biol. Chem. *386*, 1105–1114.

Eberth, A., Lundmark, R., Gremer, L., Dvorsky, R., Koessmeier, K.T., McMahon, H.T., and Ahmadian, M.R. (2009). A BAR domain-mediated autoinhibitory mechanism for RhoGAPs of the GRAF family. Biochem. J. *417*, 371–377.

Ellenbroek, S.I. and Collard, J.G. (2007). Rho GTPases: functions and association with cancer. Clin. Exp. Metastas. *24*, 657–672.

Erickson, J.W. and Cerione, R.A. (2004). Structural elements, mechanism, and evolutionary convergence of Rho protein-guanine nucleotide exchange factor complexes. Biochemistry *43*, 837–842.

Etienne-Manneville, S. and Hall, A. (2002). Rho GTPases in cell biology. Nature *420*, 629–635.

Eva, A. and Aaronson, S.A. (1985). Isolation of a new human oncogene from a diffuse B-cell lymphoma. Nature *316*, 273–275.

Fidyk, N.J. and Cerione, R.A. (2002). Understanding the catalytic mechanism of GTPase-activating proteins: demonstration of the importance of switch domain stabilization in the stimulation of GTP hydrolysis. Biochemistry *41*, 15644–15653.

Gad, A.K. and Aspenström, P. (2010). Rif proteins take to the RhoD: Rho GTPases at the crossroads of actin dynamics and membrane trafficking. Cell. Signal. *22*, 183–189.

Garavini, H., Riento, K., Phelan, J.P., McAlister, M.S., Ridley, A.J., and Keep, N.H. (2002). Crystal structure of the core domain of RhoE/Rnd3: a constitutively activated small G protein. Biochemistry *41*, 6303–6310.

Garcia-Mata, R., Boulter, E., and Burridge, K. (2011). The 'invisible hand': regulation of RHO GTPases by RHOGDIs. Nat. Rev. Mol. Cell Biol. *12*, 493–504.

Garrett, M.D., Self, A.J., van Oers, C., and Hall, A. (1989). Identification of distinct cytoplasmic targets for ras/R-ras and rho regulatory proteins. J. Biol. Chem. *264*, 10–13.

Graham, D.L., Eccleston, J.F., and Lowe, P.N. (1999). The conserved arginine in Rho-GTPase-activating protein is essential for efficient catalysis but not for complex formation with Rho·GDP and aluminum fluoride. Biochemistry *38*, 985–991.

Graham, D.L., Lowe, P.N., Grime, G.W., Marsh, M., Rittinger, K., Smerdon, S.J., Gamblin, S.J., and Eccleston, J.F. (2002). MgF(3)(-) as a transition state analog of phosphoryl transfer. Chem. Biol. *9*, 375–381.

Guo, Z., Ahmadian, M.R., and Goody, R.S. (2005). Guanine nucleotide exchange factors operate by a simple allosteric competitive mechanism. Biochemistry *44*, 15423–15429.

Han, J., Luby-Phelps, K., Das, B., Shu, X., Xia, Y., Mosteller, R.D., Krishna, U.M., Falck, J.R., White, M.A., and Broek, D. (1998). Role of substrates and products of PI 3-kinase in regulating activation of Rac-related guanosine triphosphatases by Vav. Science *279*, 558–560.

Hart, M.J., Eva, A., Evans, T., Aaronson, S.A., and Cerione, R.A. (1991). Catalysis of guanine nucleotide exchange on the CDC42Hs protein by the dbl oncogene product. Nature *354*, 311–314.

Haslam, R.J., Koide, H.B., and Hemmings, B.A. (1993). Pleckstrin domain homology. Nature *363*, 309–310.

Heasman, S.J., and Ridley, A.J. (2008). Mammalian Rho GTPases: new insights into their functions from in vivo studies. Nat. Rev. Mol. Cell Biol. *9*, 690–701.

Hoffman, G.R., and Cerione, R.A. (2002). Signaling to the Rho GTPases: networking with the DH domain. FEBS Lett. *513*, 85–91.

Hutchinson, J.P., and Eccleston, J.F. (2000). Mechanism of nucleotide release from Rho by the GDP dissociation stimulator protein. Biochemistry *39*, 11348–11359.

Jaffe, A.B., and Hall, A. (2005). Rho GTPases: biochemistry and biology. Annu. Rev. Cell Dev. Biol. *21*, 247–269.

Jaiswal, M., Dubey, B.N., Koessmeier, K.T., Gremer, L., and Ahmadian, M.R. (2012). Biochemical assays to characterize Rho GTPases. In: Rivero, R., ed., *Rho GTPases Methods and Protocols*, Springer, Berlin, pp. 37–58.

Jaiswal, M., Dvorsky, R., and Ahmadian, M.R. (2013). Deciphering the molecular and functional basis of Dbl family proteins: a novel systematic approach toward classification of selective activation of the Rho family proteins. J. Biol. Chem. *288*, 4486–4500.

Jaiswal, M., Dvorsky, R., Amin, E., Risse, S.L., Fansa, E.K., Zhang, S.C., Taha, M.S., Gauhar, A.R., Nakhaei-Rad, S., and Kordes, C., *et al.* (2014). Functional crosstalk between Ras and Rho pathways: p120RasGAP competitively inhibits the RhoGAP activity of Deleted in Liver Cancer (DLC) tumor suppressors by masking its catalytic arginine finger. J. Biol. Chem. *280*, 6839–6849.

Jaiswal, M., Gremer, L., Dvorsky, R., Haeusler, L.C., Cirstea, I.C., Uhlenbrock, K., and Ahmadian, M.R. (2011). Mechanistic insights into specificity, activity, and regulatory elements of the regulator of G-protein signaling (RGS)-containing Rho-specific guanine nucleotide exchange factors (GEFs) p115, PDZ-RhoGEF (PRG), and leukemia-associated RhoGEF (LARG). J. Biol. Chem. *286*, 18202–18212.

Jefferson, A.B. and Majerus, P.W. (1995). Properties of type II inositol polyphosphate 5-phosphatase. J. Biol. Chem. *270*, 9370–9377.

Lancaster, C.A., Taylor-Harris, P.M., Self, A.J., Brill, S., van Erp, H.E., and Hall, A. (1994). Characterization of rhoGAP. A GTPase-activating protein for rho-related small GTPases. J. Biol. Chem. *269*, 1137–1142.

Lemmon, M.A. (2004). Pleckstrin homology domains: not just for phosphoinositides. Biochem. Soc. T. *32*, 707–711.

Lemmon, M.A., Ferguson, K.M., and Abrams, C.S. (2002). Pleckstrin homology domains and the cytoskeleton. FEBS Lett. *513*, 71–76.

Ligeti, E., Dagher, M.-C., Hernandez, S.E., Koleske, A.J., and Settleman, J. (2004). Phospholipids can switch the GTPase substrate preference of a GTPase-activating protein. J. Biol. Chem. *279*, 5055–5058.

Liu, X., Wang, H., Eberstadt, M., Schnuchel, A., Olejniczak, E.T., Meadows, R.P., Schkeryantz, J.M., Janowick, D.A., Harlan, J.E., and Harris, E.A., *et al.* (1998). NMR structure and mutagenesis of the N-terminal Dbl homology domain of the nucleotide exchange factor Trio. Cell *95*, 269–277.

Minoshima, Y., Kawashima, T., Hirose, K., Tonozuka, Y., Kawajiri, A., Bao, Y.C., Deng, X., Tatsuka, M., Narumiya, S., and May, W.S., Jr., *et al.* (2003). Phosphorylation by aurora B converts MgcRacGAP to a RhoGAP during cytokinesis. Dev. Cell *4*, 549–560.

Moon, S.Y., and Zheng, Y. (2003). Rho GTPase-activating proteins in cell regulation. Trends Cell Biol. *13*, 13–22.

Nassar, N., Hoffman, G.R., Manor, D., Clardy, J.C., and Cerione, R.A. (1998). Structures of Cdc42 bound to the active and catalytically compromised forms of Cdc42GAP. Nat. Struct. Biol. *5*, 1047–1052.

Nimnual, A.S., Yatsula, B.A., and Bar-Sagi, D. (1998). Coupling of Ras and Rac guanosine triphosphatases through the Ras exchanger Sos. Science *279*, 560–563.

Rittinger, K., Taylor, W.R., Smerdon, S.J., and Gamblin, S.J. (1998). Support for shared ancestry of GAPs. Nature *392*, 448–449.

Rittinger, K., Walker, P.A., Eccleston, J.F., Nurmahomed, K., Owen, D., Laue, E., Gamblin, S.J., and Smerdon, S.J. (1997a). Crystal structure of a small G protein in complex with the GTPase-activating protein rhoGAP. Nature *388*, 693–697.

Rittinger, K., Walker, P.A., Eccleston, J.F., Smerdon, S.J., and Gamblin, S.J. (1997b). Structure at 1.65 A of RhoA and its GTPase-activating protein in complex with a transition-state analogue. Nature *389*, 758–762.

Roberts, P.J., Mitin, N., Keller, P.J., Chenette, E.J., Madigan, J.P., Currin, R.O., Cox, A.D., Wilson, O., Kirschmeier, P., and Der, C.J. (2008). Rho family GTPase modification and dependence on CAAX motif-signaled posttranslational modification. J. Biol. Chem. *283*, 25150–25163.

Rose, R., Weyand, M., Lammers, M., Ishizaki, T., Ahmadian, M.R., and Wittinghofer, A. (2005). Structural and mechanistic insights into the interaction between Rho and mammalian Dia. Nature *435*, 513–518.

Rossman, K.L., Der, C.J., and Sondek, J. (2005). GEF means go: turning on RHO GTPases with guanine nucleotide-exchange factors. Nat. Rev. Mol. Cell. Biol. *6*, 167–180.

Scheffzek, K. and Ahmadian, M.R. (2005). GTPase activating proteins: structural and functional insights 18 years after discovery. Cell. Mol. Life Sci. *62*, 3014–3038.

Scheffzek, K., Ahmadian, M.R., Kabsch, W., Wiesmuller, L., Lautwein, A., Schmitz, F., and Wittinghofer, A. (1997). The Ras-RasGAP complex: structural basis for GTPase activation and its loss in oncogenic Ras mutants. Science *277*, 333–338.

Scheffzek, K., Ahmadian, M.R., and Wittinghofer, A. (1998). GTPase-activating proteins: helping hands to complement an active site. Trends Biochem. Sci. *23*, 257–262.

Scheffzek, K., Lautwein, A., Kabsch, W., Ahmadian, M.R., and Wittinghofer, A. (1996). Crystal structure of the GTPase-activating domain of human p120GAP and implications for the interaction with Ras. Nature *384*, 591–596.

Tnimov, Z., Guo, Z., Gambin, Y., Nguyen, U.T., Wu, Y.W., Abankwa, D., Stigter, A., Collins, B.M., Waldmann, H., Goody, R.S., *et al.* (2012). Quantitative analysis of prenylated RhoA interaction with its chaperone, RhoGDI. J. Biol. Chem. *287*, 26549–26562.

Tolias, K.F., Cantley, L.C., and Carpenter, C.L. (1995). Rho family GTPases bind to phosphoinositide kinases. J. Biol. Chem. *270*, 17656–17659.

Trahey, M. and McCormick, F. (1987). A cytoplasmic protein stimulates normal N-ras p21 GTPase, but does not affect oncogenic mutants. Science *238*, 542–545.

Tyers, M., Haslam, R.J., Rachubinski, R.A., and Harley, C.B. (1989). Molecular analysis of pleckstrin: the major protein kinase C substrate of platelets. J. Cell Biochem. *40*, 133–145.

Vetter, I.R. and Wittinghofer, A. (2001). The guanine nucleotide-binding switch in three dimensions. Science *294*, 1299–1304.

Viaud, J., Gaits-Iacovoni, F., and Payrastre, B. (2012). Regulation of the DH–PH tandem of guanine nucleotide exchange factor for Rho GTPases by phosphoinositides. Adv. Biol. Reg. *52*, 303–314.

Vigil, D., Cherfils, J., Rossman, K.L., and Der, C.J. (2010). Ras superfamily GEFs and GAPs: validated and tractable targets for cancer therapy? Nat. Rev. Cancer *10*, 842–857.

Williams, C.L. (2003). The polybasic region of Ras and Rho family small GTPases: a regulator of protein interactions and membrane association and a site of nuclear localization signal sequences. Cell. Signal. *15*, 1071–1080.

Wittinghofer, A. and Vetter, I.R. (2011). Structure-function relationships of the G domain, a canonical switch motif. Annu. Rev. Biochem. *80*, 943–971.

Yamashita, T. and Tohyama, M. (2003). The p75 receptor acts as a displacement factor that releases Rho from Rho-GDI. Nat. Neurosci. *6*, 461–467.

Yang, X.Y., Guan, M., Vigil, D., Der, C.J., Lowy, D.R., and Popescu, N.C. (2009). p120Ras-GAP binds the DLC1 Rho-GAP tumor suppressor protein and inhibits its RhoA GTPase and growth-suppressing activities. Oncogene *28*, 1401–1409.

Part 2
Rho signaling in cell physiology

Rho GTPases in cadherin-based cell–cell interactions: Adhesion, repulsion, fusion

4

Fernanda Bajanca*, Elena Scarpa†, and Eric Theveneau*,†

*Centre de Biologie du Développement (UMR5547), Centre de Biologie Intégrative, Université de Toulouse, CNRS, UPS, France

†Department of Physiology, Development and Neuroscience, University of Cambridge, Cambridge, UK

†eric.theveneau@univ-tlse3.fr

Keywords: Rho, Rac, Cdc42, cadherin, adhesion, contact inhibition of locomotion, membrane fusion, neural crest, myoblast.

4.1. Introduction

Cell–cell interactions are ubiquitous in multicellular organisms. Cells assemble into organs, transiently interact, or fuse. They use a wide range of adhesion molecules to do so, among which the calcium-dependent adhesion molecules called cadherins are the most studied. The molecular networks underpinning the formation of cadherin-based adhesion have been extensively studied and the critical role of Rho GTPases has been highlighted. This family of proteins is implicated in all aspects of adhesion from assembly to clustering to recycling. However, it is sometimes difficult to understand why similar molecular machineries may lead to different, and somewhat opposite, outcomes. Here we review the current

55

knowledge about the roles of Rho GTPases in the formation of cadherin-based adhesion in vertebrates and discuss it in the context of cadherin adhesion leading to contrasting outcomes such as cell repulsion and cell fusion.

4.1.1. *Rho GTPases*

Rho GTPases belong to the Ras GTPases family. In vertebrates, there are twenty Rho GTPases classified into eight subfamilies (Fort and Theveneau, 2014; Sadok and Marshall, 2014). Two types can be distinguished: typical Rho GTPases, whose activity depends on binding to GTP, and atypical ones regulated at the transcriptional level (see Chapters 1 and 2 by Fort and Aspenström). Typical Rho GTPases are controlled by (i) guanine nucleotide exchange factors (GEFs) that help Rho GTPases exchange GDP for GTP, maintaining them active; (ii) GTPase activating proteins (GAPs) that inactivate RhoGTPases by stimulating hydrolysis of GTP into GDP; (iii) guanine nucleotide dissociation inhibitor (GDI) that bind to GDP-bound GTPases and prevent them from reloading GTP (see Chapter 3 by Amin and Ahmadian).

Above all, Rho GTPases regulate the cytoskeleton and as such participate in polarity and membrane dynamics and all associated cellular processes from migration (Sadok and Marshall, 2014) to division (Chircop, 2014) to differentiation (Hoon *et al.*, 2016). Rho GTPases also modulate signaling pathways and go to the nucleus, contributing to transcriptional regulation. These aspects of Rho biology are not covered by this chapter, but readers can find publications on these topics (Bachmann *et al.*, 2013; Schlessinger *et al.*, 2009; Sebe *et al.*, 2010).

4.1.2. *Cadherins*

Vertebrate cells establish various types of cell–cell junctions: (i) tight junctions, (ii) adherens junctions, (iii) gap junctions, and (iv) desmosomes (Giepmans and van Ijzendoorn, 2009). In this chapter, we focus solely on adherens junctions which are made of cadherins.

Cadherins are calcium-dependent adhesion molecules. There are over a hundred cadherins identified in humans, organized in: (i) classical cadherins (CDHs), (ii) protocadherin (PCDH), and (iii) unconventional, ungrouped, cadherin-related (CDHR) molecules (http://www.genenames.org/cgi-bin/genefamilies/set/16). We focus on type I (i.e., cadherins E, N, C, and P) and type II (i.e., cadherin-7, 11 and VE-cadherin) molecules from the CDH family. Both type I and II have a single-pass transmembrane region and a conserved cytoplasmic domain containing binding motifs for armadillo domain proteins of the catenin family. P120 catenin interacts with the juxta membrane domain of cadherins, while β catenin interacts with the C-terminal domain and links cadherins to actin via the vinculin homolog α catenin (Nelson, 2008). Cadherins cluster in *cis* (within the same cell) and *trans* (with cadherins from an adjacent cell) to form adherens junctions. In contrast with type I cadherins, type II have not been reported to form *cis* lateral interactions (Brasch *et al.*, 2011; Patel *et al.*, 2006), which may account for their diminished strength of cell–cell adhesion. Adherens junctions bridge the actin cytoskeleton of adjacent cells. They confer rigidity to a cell aggregate and thus help maintain tissue integrity in response to mechanical stress.

4.2. Forming a cadherin-based epithelial adhesion

When two cells contact each other, one of the possible outcomes is the formation of adherens junctions (Ratheesh *et al.*, 2013). Most of the data in vertebrates come from studies on normal or cancerous cell lines derived from breasts, kidneys, skin, or blood vessels obtained from humans, dogs, and mice. The cellular background has an influence on the formation and dynamics of adherens junctions. As such, data obtained on a given line may not be transferable to another. In addition, most studies have focused on E-cadherin in epithelial cells and VE-cadherin in endothelial cells, and whether the conclusions can be extended to all cadherins remains unclear. Further, the type of matrix on which cells are cultured or the stage of adherens junctions' life studied influences the experiments and their conclusions. Thus, it is difficult to reconcile all data on Rho GTPases in cadherin-based adhesions. Here, we tried to summarize accepted data about epithelial junction formation to compare them with cell repulsion or fusion.

4.2.1. Forming a junction

In epithelial cells, the formation of an adherens junction can be summarized as follows: (i) cells touch due to cytoskeleton-driven membrane activity (ruffling, protrusions, cell displacement); (ii) cadherins interact, adhesion strength increases due to linkage to the actin cytoskeleton; (iii) around the adhesion site, actin polymerization favors membrane over/underlapping while myosin-driven contraction promotes contact expansion (Fig. 1(a)). Contact size increases laterally from the initial site of contact (Fig. 1(b)). Cadherin clustering occurs at the centre of the junction. The cadherin core is surrounded by a zone of actin polymerization which is in turn surrounded by an area of actomyosin contraction (Fig. 1(c)).

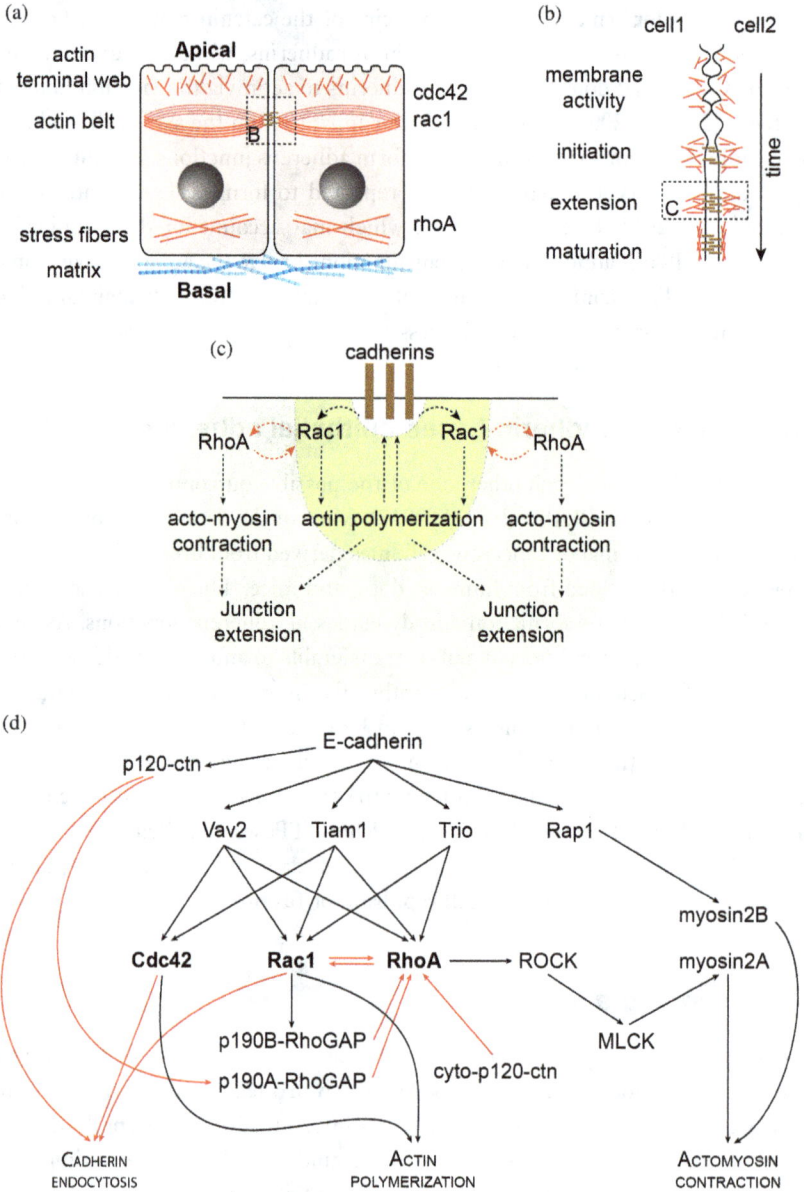

Figure 1. Cadherins and Rho GTPases in epithelial junctions (see text for details).

4.2.2. *Rho GTPases in adherens junction formation*

Cadherin engagement triggers a transient peak of Rac1 activity around the cadherin clusters, as shown for E-, M-, and VE-cadherin (Kitt and Nelson, 2011; Kovacs *et al.*, 2002; Noren *et al.*, 2001; Perez *et al.*, 2008). The effect of N-cadherin on Rac1 activity in epithelial junction formation was not studied while N-cadherin mediates Rac1 inhibition during collisions or fusion (Charrasse *et al.*, 2002; Theveneau *et al.*, 2010).

Glial cells and neurons rely on N-cadherin to crawl on each other suggesting that Rac1-driven actin polymerization and N-cadherin-based junctions could locally coexist; however, in this case, the protrusive activity is RhoA (not Rac1!) dependent (Shikanai *et al.*, 2011; Xu *et al.*, 2015). It would be interesting to assess Rac1 dynamics during junctions' formation in epithelial cells that rely on N-cadherin for adhesion such as neural tube or ovary cells.

The local Rac1 hotspot observed around the initial cadherin cluster is hedged in by a peak of RhoA activity (Fig. 1(c)). In this simplified view, an assembling junction is a core of clustered cadherins with two concentric rings at its edge: an inner one of Rac1-driven membrane ruffles and an outer one of RhoA-dependent actomyosin contraction (Yamada and Nelson, 2007). Both contribute to cadherin recruitment and clustering driving expansion. The Rac and Rho areas do not mix due to mutual inhibition between the two GTPases (Burridge and Wennerberg, 2004; Sander *et al.*, 1999).

4.2.3. *Rho and Rac during junction assembly*

RhoA and Rac1 act at the membrane. Therefore, controlling their recruitment is essential. Below, we discuss some examples of how Rac and Rho activities are coordinated via common cofactors (Fig. 1(d)).

Rho GTPases' activity depends on the stoichiometry and localization of GEFs and GAPs. Several GEFs such as Tiam1 (Hordijk *et al.*, 1997; Malliri *et al.*, 2004), Trio (Timmerman *et al.*, 2015; Tu and You, 2014), Vav2, and the GTPase Rap1 (Fukuyama *et al.*, 2006; Kooistra *et al.*, 2007) are found at junctions together with GAPs including Rho GAPs p190A (ARHGAP35) (Noren *et al.*, 2003; Wildenberg *et al.*, 2006) and p190B (ARHGAP5) (Bustos *et al.*, 2008). Vav2 can activate Rac1, RhoA, and Cdc42. Tiam1 is primarily a Rac1/Cdc42 activator with weak effects on RhoA. Trio can activate Rac1 and RhoA. Rap1 is an activator of myosin-2B and vav2 whereas p190A and p190B are inhibitors of RhoA. When cadherins interact, local recruitment of Tiam1 and Vav2 activates Rac1. Rac1, in turn, activates p190B-RhoGAP (Citi *et al.*, 2014; Ratheesh *et al.*, 2013). Therefore, the local activation of Rac1 directly feeds back into RhoA inhibition.

Cadherins also recruit p120-catenin which recruits p190A-RhoGAP (Wildenberg *et al.*, 2006; Zebda *et al.*, 2013), thus amplifying local inhibition of RhoA. Cytoplasmic p120-catenin has a negative effect on RhoA activity probably by trapping RhoA-GDP (Anastasiadis *et al.*, 2000; Anastasiadis and Reynolds, 2000) suggesting that high local concentration of p120 will further inhibit RhoA by reducing the pool of RhoA-GDP available for reloading of GTP. Further, binding of p120 to cadherins masks endocytic signals preventing cadherin endocytosis (Nanes *et al.*, 2012) and stabilizing junctions (Kourtidis *et al.*, 2013).

Cdc42, together with Rac1, favours actin polymerization and reduces cadherin endocytosis at junctions (Izumi *et al.*, 2004). While epithelial junctions assemble, apicobasal polarity sets in. Cdc42 lies upstream of Par6 and aPKC, two members of the PAR complex, essential for apical identity (Mack and Georgiou, 2014). Local Cdc42 activity is important for maturation of the apical domain and is required for further coupling of cells by tight junctions (Citi *et al.*, 2014; Mack and Georgiou, 2014).

Due to activation of Rac1 by cadherins and junction stabilization by Rac1 and Cdc42, most of RhoA activity is detected on the junction's periphery during assembly. RhoA triggers a cascade involving ROCK (Rho Kinase) and myosin light chain kinase that activates myosin-2A. This contraction at the junction favours cadherin clustering and expansion. myosin-2B lies downstream of Rap1 and contributes to Rho-dependent contractility (Ratheesh *et al.*, 2013).

Not all myosins positively participate in Rho signaling. Myosin-9A for instance is known to act as a GAP for RhoA promoting conversion of RhoA-GTP into RhoA-GDP. Blocking myosin-9A can lead to epithelial defects due to an increase of RhoA and a subsequent partial loss of E-cadherin (Abouhamed *et al.*, 2009) or disrupt the formation of actin bundles observed after contact initiation (Omelchenko and Hall, 2012).

Adherens junctions interact with the actin cytoskeleton via β- and α-catenins. Rac1/Cdc42-controlled actin dynamics, RhoA/myosin-dependent contractility and catenin-dependent attachment to cadherin generate tension through actin cables. Tension favors junction assembly and stability and modulates recruitment of cofactors such as α-actinin and vinculin (Choi *et al.*, 2016; Ishiyama *et al.*, 2013; Leckband and de Rooij, 2014; Liu *et al.*, 2010).

During junction maturation, the organization of the actin network changes from branched cables perpendicular to the membrane to actin bundles parallel to the membrane. To a lesser extent, adherens junctions also interact with the microtubule network. During initiation, microtubules favor myosin-2B accumulation at junctions and microtubules plus-end dynamics is essential for cadherin clustering (Stehbens *et al.*, 2006). Also, in mature junctions, microtubules, via the centralspindlin tetramer, recruit Ect2 that acts as a GEF for RhoA and inhibits Rac1 (Ratheesh *et al.*, 2012). Contrary to what is observed during formation, in stabilized junctions, Rac and Rho activities are codetected and equally important for junction stability (Citi *et al.*, 2014).

For the roles of other GEFs, GAPs, binding partners, or signaling pathways involved, see Citi *et al.* (2014), Nola *et al.* (2011), Ratheesh *et al.* (2013), and Terry *et al.* (2011).

4.2.4. *Cross talk between adherens junctions and focal adhesions*

The formation and turnover of cell–matrix adhesions is regulated by Rho GTPases (Dubash *et al.*, 2009; Kanchanawong *et al.*, 2010), and polarized epithelial cells maintain their cell–cell and cell–matrix adhesions in separate domains (Burute and Thery, 2012). Competition for the regulation of Rho GTPases could contribute to this antagonism between focal adhesion and adherens junctions. Focal adhesion kinase (FAK) and paxillin are important effectors of focal adhesions and often contribute to Rac-dependent pathways. However, FAK and paxillin can also inhibit Rac1 (Yano *et al.*, 2004). This effect was observed in cells that form N-cadherin-based junctions, and decreasing Rac1 activity helped junctions to form (Yano *et al.*, 2004). In cells expressing cadherins that trigger Rac1 activity, like E-cadherin, junction initiation is likely to be impaired by the proximity of focal adhesion signaling. In addition, FAK can phosphorylate β-catenin (Chen *et al.*, 2012) and thus promote detachment of cadherins from microfilaments. Also, cell–matrix signaling via Syndecans can influence the composition of adherens junctions (Gopal *et al.*, 2016).

Another line of cross talk between focal adhesion and adherens junction is tension. Indeed, both structures are under tension (Mui *et al.*, 2016). Adherens junctions undergo mechanically induced growth in response to loading, via Rac1 and Myosin (Liu *et al.*, 2010). Differential response to tension could be even more relevant if cells are in movement. It may be mechanically impossible for cells to sustain a stable cadherin junction near cell–matrix interaction. Spatial segregation between cadherins and integrins might occur as a consequence of minimization of tensional forces in a tissue (Tseng *et al.*, 2012).

4.3. Repulsion: Contact inhibition of locomotion

Cells do not always establish a stable junction. A contact can indeed be followed by a rapid detachment. One of the best examples of this behaviour is contact inhibition of locomotion in which cells cease migrating in response to a contact with another cell (Mayor and Carmona-Fontaine, 2010). The contact-inhibition behaviour can be summarized as follows: (i) cells make a local contact, (ii) protrusions are inhibited at the contact, (iii) a new polarity emerges with the cell's rear at the contact and the cell's front at the cell-free edge, (iv) cells produce new protrusions toward the cell-free space, (v) cells disassemble their junctions, and (vi) cells resume migration (Fig. 2).

4.3.1. *Xenopus neural crest cells as a model for contact inhibition of locomotion*

As an example of contact-inhibition behaviour, we discuss data on Xenopus cranial neural crest cells. The neural crest is an embryonic cell population emerging from the dorsal neuroepithelium during neurulation (Theveneau and Mayor, 2012). These cells are initially epithelial and progressively start emigrating as mesenchymal cells. Then, neural crest cells undergo an extensive migration to colonize the whole embryo. The premigratory epithelial phenotype of neural crest cells is mediated by stable E-cadherin junctions. When cells convert to a mesenchymal phenotype, they switch to N-cadherin/cadherin-11 transient cell–cell interactions (Taneyhill and Schiffmacher, 2017) (Fig. 2(a)). E-cadherin remains expressed but is no longer involved in cell–cell adhesion (Huang *et al.*, 2016). On the contrary, N-cadherin or cadherin-11 knockdown abolishes migratory neural crest cells' ability to interact (Becker *et al.*, 2013; Kuriyama *et al.*, 2014; Theveneau *et al.*, 2010).

4.3.2. *Rho GTPases in contact inhibition between Xenopus neural crest cells*

When two neural crest cells collide, they exhibit a contact-inhibition response (Stramer and Mayor, 2016) (Fig. 2(b)). We can summarize the role of cadherins and Rho GTPases during contact inhibition of locomotion in *Xenopus* neural crest cells as follows. Two Rac1-positive protrusions collide, and neural crest cells establish a contact (Figs. 2(c) and 2(d)). This contact involves N-cadherin and cadherin-11 and triggers the noncanonical Wnt planar cell polarity pathway. N-cadherin inhibits Rac1 while the Wnt/PCP pathway activates RhoA (Fig. 2(e)). How are the contact-dependent decrease of Rac1 and the concomitant induction of RhoA orchestrated? How is Rac activity redirected on the opposite side of the cell?

It was shown in cell lines endogenously expressing N-cadherin that cytoplasmic p120 can activate Rac1 and Cdc42 promoting protrusive activity

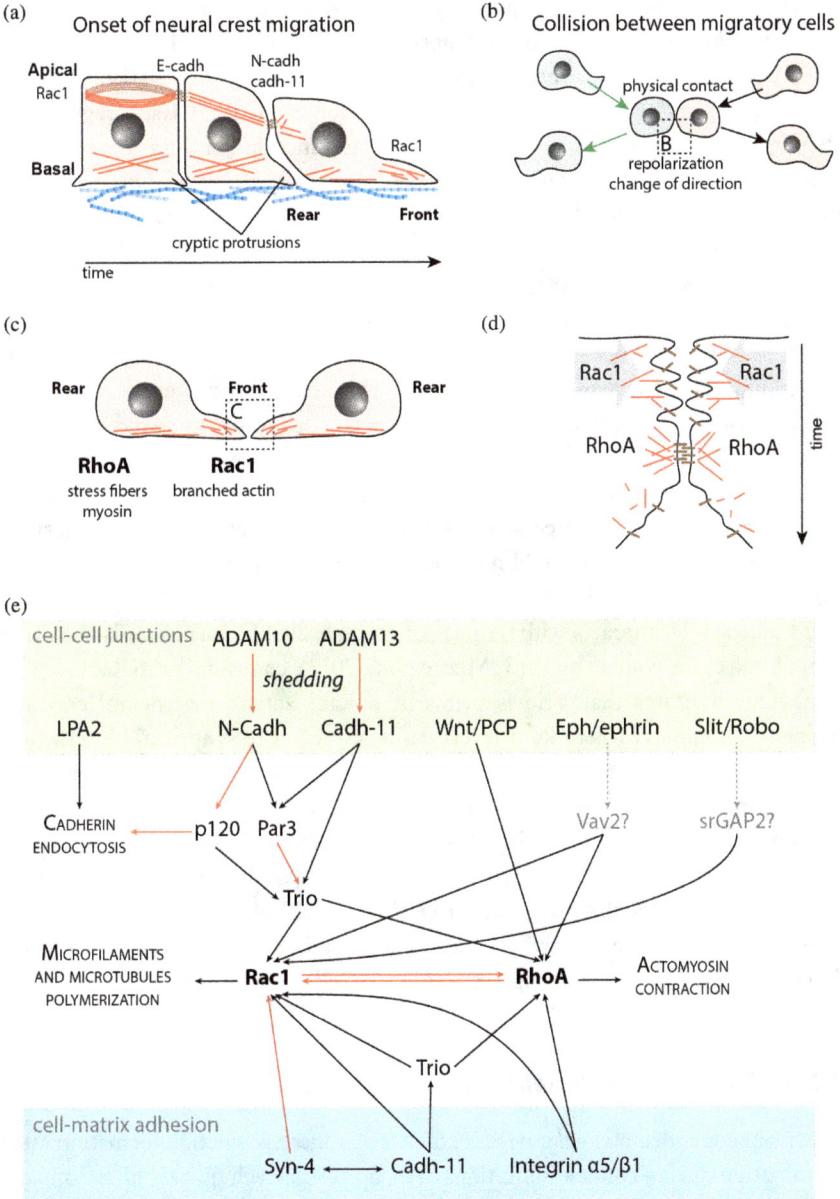

Figure 2.   Cadherins and Rho GTPases in contact inhibition of locomotion (see text for details).

(Grosheva *et al.*, 2001). Overexpressing E-cadherin in these cells sequesters p120 at the junction, and its effect on cell morphology is abolished. This indicates that cell–matrix and cell–cell adhesions can compete for binding to p120 (Boguslavsky *et al.*, 2007) and that this competition contributes to local Rac1 regulation. Premigratory neural crest cells form E-cadherin junctions, and Rac1 is activated at the junctions in a p120-dependent manner (Scarpa *et al.*, 2015). During migration, E-cadherin is needed for motility but no longer required for cell–cell adhesion (Huang *et al.*, 2016), and neural crest cell–cell adhesion relies on N-cadherin instead (Kuriyama *et al.*, 2014; Theveneau *et al.*, 2010). Thus, in neural crest, the switch for E- to N-cadherin-based junctions might free some p120, allowing Rac1 activation outside of cell–cell junctions. Alternatively, p120 may inhibit Rac1 downstream of N-cadherin by counteracting integrin-dependent activation in the vicinity of the cell–cell contact, as suggested by a study in mouse fibroblasts (Ouyang *et al.*, 2013). During heterotypic N-cadherin-dependent contact inhibition between placodes and neural crest cells, focal adhesions are disassembled at the site of cell–cell contact (Theveneau *et al.*, 2013) suggesting that, in neural crest, p120 could mediate the local inhibition of cell–matrix adhesion downstream of N-cadherin.

Cadherin-11 interacts with Trio (Kashef *et al.*, 2009). During collisions, Trio is inhibited at the contact by Par3 (Moore *et al.*, 2013), contributing to Rac1 inhibition. This indicates that Trio is primarily a Rac1 activator in neural crest. It happens in a context where Rac1 levels are kept low due to syndecan-4 interaction with the matrix (Matthews *et al.*, 2008). Blocking syndecan-4 leads to an increase of Rac1 and prevents cells from polarizing.

Other pathways modulating RhoA (EphA/ephrinA/vav2 (Batson *et al.*, 2014)) and Rac1 (slit/Robo/srGAP2 (Fritz *et al.*, 2015)) during contact inhibition were studied in fibroblasts and cancer cells and could play a role in neural crest cells since some Robo (Hocking *et al.*, 2010) and Eph/ephrins (Helbling *et al.*, 1999; Smith *et al.*, 1997) are expressed by these cells.

4.3.3. *Releasing adhesion: tension, forces*

The main conundrum is what triggers failure of adherens junction formation after its initiation during contact inhibition? Why do cells detach quickly after contact? Is detachment related to the change of polarity? There are several hypotheses including (i) mechanical signaling, (ii) mechanical force, (iii) competition for common effectors between cell–cell and cell–matrix adhesions, and (iv) inhibition of cadherin.

Cadherins dynamically interact with the cytoskeleton, and RhoA is activated at cell–cell contacts during contact inhibition. Therefore, contractility could be responsible for mechanically pulling membranes away. Rho/ROCK signaling is indeed required for contact inhibition (Carmona-Fontaine *et al.*, 2008; Kadir *et al.*, 2011), but myosin inhibition suggests that contractility is partially dispensable for contact inhibition in fibroblasts (Kadir *et al.*, 2011) and neural crest cells (Scarpa, unpublished observation). During epithelial contact formation, such Rho-dependent contraction contributes to cadherin clustering and contact enlargement. There is an important difference. In epithelial junction, the actin network orientation switches from perpendicular to parallel to the contact, which is not observed during contact inhibition in neural crest cells (Scarpa *et al.*, 2015). Thus, in neural crest, tension pulls cadherins apart, whereas in epithelial junctions, it brings cadherins together. The fact that contact inhibition still happens under myosin inhibition suggests that other forces may contribute to pull membranes apart.

That force could come from the new lamellipodia at the cell's free edge. Preventing lamellipodia formation can inhibit contact disassembly (Scarpa *et al.*, 2015). Conversely, activating Rac1 at the free edge in cells that are not supposed to dissociate promotes disassembly. Yet, nothing in these experiments indicates that the force generated by protrusive activity is responsible for the inhibition of contacts. In fact, contact disassembly during physiological contact-inhibition response is shorter than with experimentally induced lamellipodia. The influence of tension on cell polarity has also been tested in *Xenopus* mesodermal cells (Weber *et al.*, 2012), and a similar delay was observed between applied tension and reversal of polarity. Similar feedbacks between polarity and tensions are well described in migratory epithelial sheets (Trepat and Fredberg, 2011). While tension from protrusions likely stabilizes the contact-induced polarity and in turn contributes to junction inhibition, reasonable doubts remain concerning the direct role of the protrusive force in physically separating cells.

An alternative to a transmitted mechanical signal from front to rear is pure mechanical force breaking the rear of the cell without affecting junctions as observed in chick neural crest (Ahlstrom and Erickson, 2009). In such case, contact disassembly and protrusive activity need not be linked by signaling.

Another option is membrane tension: it was shown in neutrophils that plasma membrane tension is essential for maintaining Rac1 polarization confined on one side of the cell (Houk *et al.*, 2012). Because tension is transmitted instantaneously, a small difference in the level of Rac1 activity between free edge and contact might be amplified by feedback loops sensing the tension (Diz-Munoz *et al.*, 2016; Tsujita *et al.*, 2015).

4.3.4. *Releasing adhesion: competition for cofactors and direct inhibition of cadherins*

Focal adhesions and adherens junctions may compete for common effectors (i.e, actin, vinculin). Forming a lamellipodium could decrease the G-actin pool and favor F-actin catastrophe. Indeed, induction of protrusions can lower total G-actin levels by 40% within 4 minutes (Kiuchi *et al.*, 2011), a time-scale compatible with the dynamics of the contact-inhibition response. Thus, formation of a protrusion could affect cadherin adhesion by impairing F-actin turnover at the site of contact. The same competition effect could also drive the local detachment from the substrate observed directly underneath the point of contact during collision (Theveneau *et al.*, 2013). Yet, it is not clear how Rac1 activity driving protrusion formation rapidly increases away from the contact. This could also be due to local catastrophe of actin filaments and focal-adhesion disassembly since Rac1 associated with these two structures diffuses less (Hinde *et al.*, 2013; Lakhani *et al.*, 2015). Thus, increased diffusion of Rac1 could contribute to the polarity switch observed after contact.

Another possibility is that specific Rac1 regulators act differentially on the cell–cell contact and free-edge Rac1 pool. During contact inhibition in fibroblasts, the Rac-GAP srGAP2 dampens Rac1 activity at the contact, leaving Rac1 at the lamellipodium unaffected (Fritz *et al.*, 2015). A similar regulation may occur in neural crest.

Finally, cells could separate due to direct inhibition of cadherins by endocytosis or shedding. Preventing N-cadherin endocytosis is enough to block cell dispersion without affecting motility (Kuriyama *et al.*, 2014). In addition, *Xenopus* neural crest cells express numerous proteases (Alfandari *et al.*, 1997; Christian *et al.*, 2013; Neuner *et al.*, 2009). ADAM10 and ADAM13, respectively known for shedding N-cadherin (Shoval *et al.*, 2007) and cadherin-11 (Abbruzzese *et al.*, 2016), are expressed during *Xenopus* cephalic neural crest migration and could be involved in contact release. Cleavage of extracellular domain of cadherins and associated cytoplasmic fragments could modulate Rho GTPases due to the interaction of N-cadherin and cadherin-11 with p120-catenin, Trio and Par3. Whether contact localization and activity of these enzymes is regulated by cell polarity or protrusive activity remains to be assessed.

4.4. Fusion of muscle cell membrane

Most cells are mononucleate, but a number of cell types fuse into multinucleate cells as part of their normal differentiation process. This process must be tightly regulated, and restricted to a subset of cell types, as aberrant cell fusion can result in cancer (Aguilar *et al.*, 2013; Bastida-Ruiz *et al.*, 2016; Mohr *et al.*, 2015). As an example of cell type undergoing fusion, we will focus on vertebrate myoblasts.

4.4.1. Overview of vertebrate myoblast fusion

Myogenesis is a multistep process leading to muscle formation. Proliferating myoblasts exit the cell cycle, elongate, and fuse initially with each other forming myotubes, and also with previously formed myotubes, which mature into myofibers. Myogenesis in vertebrates has been extensively studied (Biressi *et al.*, 2007; Bryson-Richardson and Currie, 2008; Buckingham, 2001, 2006; McLennan and Koishi, 2002; Thorsteinsdottir *et al.*, 2011).

Of note, *Drosophila* has been a major model to study fusion during myogenesis. Differences between *Drosophila* and vertebrates have been reviewed elsewhere (Abmayr and Pavlath, 2012; Hindi *et al.*, 2013; Kim *et al.*, 2015; Richardson *et al.*, 2008; Rochlin *et al.*, 2010). The main events of myoblast fusion are: (i) cell recognition and adhesion (Fig. 3(a)); (ii) cytoskeletal remodeling and vesicle trafficking (Fig. 3(b)); (iii) formation and expansion of fusion pores (Fig. 3(b)). Rho GTPases have been identified as critical regulators at several stages (Fig. 3(c)).

4.4.2. Rho GTPases in myoblasts fusion

Myoblasts adopt a bipolar shape before fusion, induced by actin interaction with myosin-2A at the plasma membrane (Duan and Gallagher, 2009; Nowak *et al.*, 2009; Swailes *et al.*, 2006). Just before fusion, actin is organised into an actin wall, underneath the membrane (Duan and Gallagher, 2009; Swailes *et al.*, 2004) (Figs. 3(a) and 3(b)). The microfilaments in these sheets show mixed polarity while focal adhesions are restricted to the myoblasts' tips (Peckham, 2008; Swailes *et al.*, 2004). Then, vesicles accumulate within gaps of the actin wall and fusion pores start forming (Duan and Gallagher, 2009) (Fig. 3(b)). The two cells progressively fuse into one. The newly added nucleus moves rapidly toward the central myotube nuclei (Cadot *et al.*, 2012; Duan and Gallagher, 2009).

Early *in vitro* studies demonstrated that RhoA activation by N-cadherin is required for myogenesis, while Rac1 and Cdc42 seem to play a negative role (Charrasse *et al.*, 2002; Meriane *et al.*, 2000; Wei *et al.*, 1998). However, active RhoA, through ROCK, reduces the stability and perturbs the localization of M-cadherin, which has an essential role in fusion (Charrasse *et al.*, 2006). Therefore, RhoA is rapidly deactivated after myogenesis induction, before

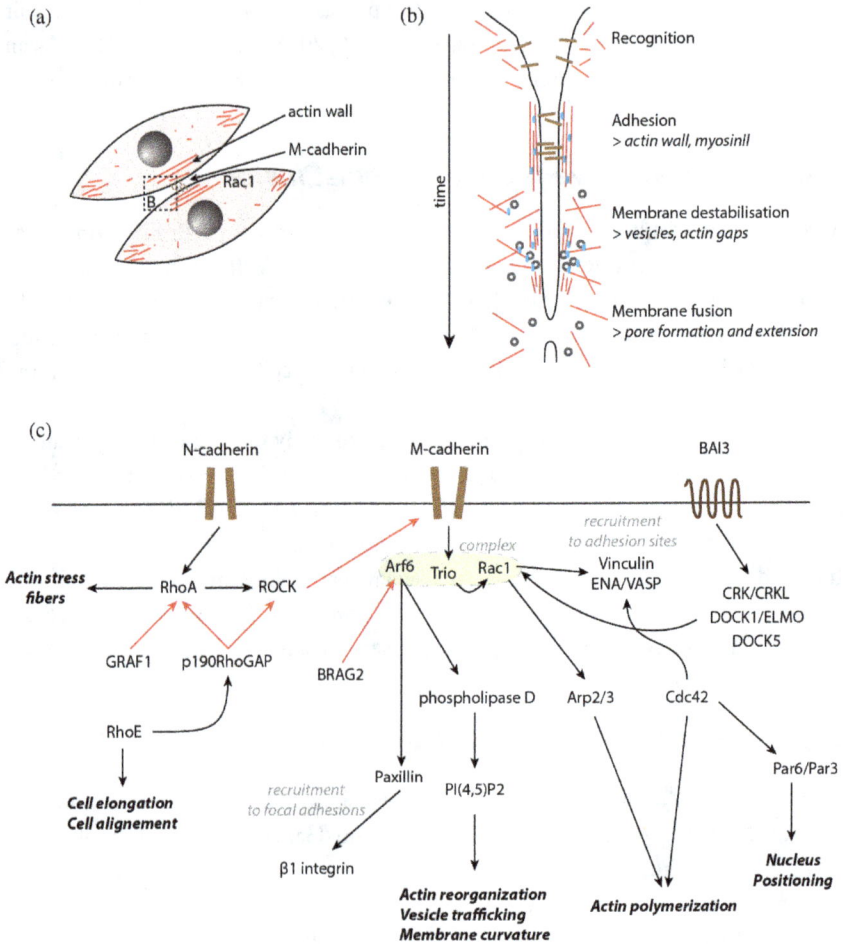

Figure 3. Rho GTPases in myoblasts fusion (see text for details).

myoblast fusion (Charrasse *et al.*, 2006; Meriane *et al.*, 2000; Nishiyama *et al.*, 2004). Two factors have been implicated: RhoE and the RhoGAP GRAF1 (GTPase regulator associated with focal adhesion kinase-1) (Doherty *et al.*, 2011; Fortier *et al.*, 2008; Lenhart *et al.*, 2014). RhoE binds and activates p190rhoGAP, which in turn inhibits RhoA and ROCK (Fortier *et al.*, 2008). Myoblast elongation and cell alignment are promoted by RhoE while the reduction of active RhoA and ROCK inhibits actin stress fibres, a hallmark of differentiating myoblasts (Duan and Gallagher, 2009; Fortier *et al.*, 2008). GRAF1 redistributes from the perinuclear region to the tips in prefused bipolar myoblasts to limit Rho-dependent actin remodeling at these sites (Doherty *et al.*, 2011). In addition, RhoA activity was found increased in GRAF1-depleted *Xenopus laevis* embryos resulting in muscle degeneration, defective motility, and embryonic lethality (Doherty *et al.*, 2011). In mice myoblasts, GRAF1 mediates RhoA down-regulation and induces muscle differentiation, while forcing expression in predifferentiated myoblasts drives robust muscle fusion by a process that requires actin remodeling (Doherty *et al.*, 2011). GRAF1-deficient mice show underdeveloped muscles and impaired regenerative capacity (Lenhart *et al.*, 2014). Isolated myoblasts depleted of GRAF1 or GRAF2 showed fusion defects due to perturbed vesicle-mediated translocation of the fusogenic ferlin proteins to the plasma membrane (Lenhart *et al.*, 2014). Therefore, regulating RhoA is probably essential to limit actin polymerization after the cell becomes fusion competent, in order to allow for the actin remodeling required for vesicle trafficking and further fusion events to proceed.

Once the fusion is initiated, Rac1 and Cdc42 activities become essential. In zebrafish, blocking Rac1 compromises fusion while constitutive Rac activation leads to hyperfusion (Srinivas *et al.*, 2007). Moreover, in mice, conditional Rac1 or Cdc42 mutagenesis reduces the fusion index both *in vivo* and *in vitro*, and the number of nuclei in developing muscle (Vasyutina *et al.*, 2009). Both Rac1 and Cdc42 control filamentous actin assembly and fusion in murine myoblasts. In addition, accumulation of vinculin, Ena-Vasp, and polymerized actin at the contact sites was markedly reduced in Rac1- and Cdc42-deficient myoblasts. However, recruitment of the actin polymerization-inducing complex Arp2/3 to the contact sites was affected only in Rac1 mutant cells, revealing a nonredundant role, or a compensation effect, for Cdc42 (Vasyutina *et al.*, 2009). One possibility is that Cdc42 plays a more important role in a later phase, when the myoblast nuclei move into the new myofiber after fusion. The movement of the myoblast nucleus toward the central myotube nuclei is driven by microtubules and dynein/dynactin complex and requires Cdc42 and its effector Par6 interacting with Par3 (Cadot *et al.*, 2012).

While the role of Cdc42 in fusion is not fully understood, we have a more complete picture upstream and downstream of Rac1. GEFs such as Dock1, Dock5, and Trio regulate Rac1 during fusion to promote actin reorganization contributing to vesicle trafficking and membrane fusion. The first step toward fusion is recognition and adhesion. In vertebrates, the receptor BAI3 (G-protein coupled receptor brain-specific angiogenesis inhibitor) (Hamoud *et al.*, 2014; Hochreiter-Hufford *et al.*, 2013) engages the ELMO/DOCK1 pathway, which activates Rac1, to mediate fusion (Hamoud *et al.*, 2014). Fusion is abolished in BAI3 deficient myoblasts, while interfering *in vivo* with BAI3–ELMO interaction impairs fusion (Hamoud *et al.*, 2014). While DOCK1 is the main GEF in vertebrate myoblast fusion, DOCK5 and the Dock-adaptor proteins Crk and Crk-like 1 (Crkl) were also implicated in *in vivo* studies in mice and/or zebrafish (Laurin *et al.*, 2008; Moore *et al.*, 2007).

Upon initial contact, M-cadherin adhesion enhances cell–cell proximity and triggers further signaling required to proceed to fusion. M-cadherin activates Rac1 via Trio (Charrasse *et al.*, 2007). A complex composed of M-cadherin, Rac1, Trio, and Arf6 forms (Bach *et al.*, 2010). Phospholipase D was identified as a downstream target of Arf6, whose activation leads to production of phosphatidylinositol 4,5-bisphosphate [PI(4,5)P2] (Bach *et al.*, 2010). This pathway may be responsible for regulating actin reorganization at the membrane, vesicle trafficking, membrane curvature and recycling (D'Souza-Schorey and Chavrier, 2006; Donaldson, 2008). Moreover, Arf6, activated by BRAG2 and Dock180 (Dock1), is required for transporting paxillin to focal adhesions, where it complexes with integrins thereby maintaining their structural integrity, which is required during myotube maturation (Pajcini *et al.*, 2008). This positions Rho GTPases as regulators of integrin signaling during myoblast fusion by promoting the assembly of their intracellular partners. The role of integrins in fusion can be found summarized elsewhere (Hindi *et al.*, 2013; Thorsteinsdottir *et al.*, 2011).

Interestingly, primary skeletal myofibers were relatively normal in Trio-null mice embryos, while secondary myogenesis is severely affected (O'Brien *et al.*, 2000). This data reinforces the complexity of *in vivo* myogenesis in mammals, and suggests that fusion occurring during primary and secondary myogenesis may have different requirements, relying respectively on Dock or Trio for Rac1 activation and subsequent actin remodeling (Hamoud *et al.*, 2014; Laurin *et al.*, 2008; O'Brien *et al.*, 2000).

4.5. Concluding remarks

With the current knowledge on epithelial adhesion, contact inhibition of locomotion and cell fusion, it is not clear why cells eventually adhere to one another, move away upon contact, or merge. At first glance, actors seem to be the same. Confusingly, closely related proteins like E- and N-cadherins display similar affinities for their main cofactors but seem to regulate Rac1 in an opposite manner. The solution may lie in the order of appearance of the different actors within the frame. In epithelial junction, microtubules are late players modulating cadherin turnover, apical identity, and junction stability. While, in contact inhibition, microtubules are part of the early response upon contact. Another putative explanation for the observed differences between these three contrasting outcomes is the changes in actin organization. Contact-inhibited cells fail to reorganize actin cables into parallel bundles. During fusion, parallel bundles form only transiently. Therefore, the local balance between branching and bundling modulators, such as Arp2/3 and formin, and contractility effectors like myosins is likely to be essential to drive adhesion, repulsion, or fusion.

Acknowledgements

Work in Eric Theveneau's lab is supported by the CNRS, FRM (AJE201224), Région Midi-Pyrénées and Toulouse Cancer Santé.

References

Abbruzzese, G., Becker, S.F., Kashef, J., and Alfandari, D. (2016). ADAM13 cleavage of cadherin-11 promotes CNC migration independently of the homophilic binding site. Dev. Biol. *415*, 383–390.

Abmayr, S.M. and Pavlath, G.K. (2012). Myoblast fusion: lessons from flies and mice. Development *139*, 641–656.

Abouhamed, M., Grobe, K., San, I.V., Thelen, S., Honnert, U., Balda, M.S., Matter, K., and Bahler, M. (2009). Myosin IXa regulates epithelial differentiation and its deficiency results in hydrocephalus. Mol. Biol. Cell *20*, 5074–5085.

Aguilar, P.S., Baylies, M.K., Fleissner, A., Helming, L., Inoue, N., Podbilewicz, B., Wang, H., and Wong, M. (2013). Genetic basis of cell-cell fusion mechanisms. Trends Genet. *29*, 427–437.

Ahlstrom, J.D. and Erickson, C.A. (2009). The neural crest epithelial-mesenchymal transition in 4D: a 'tail' of multiple non-obligatory cellular mechanisms. Development *136*, 1801–1812.

Alfandari, D., Wolfsberg, T.G., White, J.M., and DeSimone, D.W. (1997). ADAM 13: a novel ADAM expressed in somitic mesoderm and neural crest cells during *Xenopus laevis* development. Dev. Biol. *182*, 314–330.

Anastasiadis, P.Z. Moon, S.Y., Thoreson, M.A., Mariner, D.J., Crawford, H.C., Zheng, Y., and Reynolds, A.B. (2000). Inhibition of RhoA by p120 catenin. Nat. Cell Biol. *2*, 637–644.

Anastasiadis, P.Z. and Reynolds, A.B. (2000). The p120 catenin family: complex roles in adhesion, signaling and cancer. J. Cell Sci. *113* (*Pt 8*), 1319–1334.

Bach, A.S., Enjalbert, S., Comunale, F., Bodin, S., Vitale, N., Charrasse, S., and Gauthier-Rouviere, C. (2010). ADP-ribosylation factor 6 regulates mammalian myoblast fusion through phospholipase D1 and phosphatidylinositol 4,5-bisphosphate signaling pathways. Mol. Biol. Cell *21*, 2412–2424.

Bachmann, V.A., Bister, K., and Stefan, E. (2013). Interplay of PKA and Rac: fine-tuning of Rac localization and signaling. Small GTPases *4*, 247–251.

Bastida-Ruiz, D., Van Hoesen, K., and Cohen, M. (2016). The dark side of cell fusion. Int. J. Mol. Sci. *17*, 638.

Batson, J., Maccarthy-Morrogh, L., Archer, A., Tanton, H., and Nobes, C.D. (2014). EphA receptors regulate prostate cancer cell dissemination through Vav2-RhoA mediated cell-cell repulsion. Biol. Open *3*, 453–462.

Becker, S.F., Mayor, R., and Kashef, J. (2013). Cadherin-11 mediates contact inhibition of locomotion during *Xenopus* neural crest cell migration. PLoS One *8*, e85717.

Biressi, S., Molinaro, M., and Cossu, G. (2007). Cellular heterogeneity during vertebrate skeletal muscle development. Dev. Biol. *308*, 281–293.

Boguslavsky, S., Grosheva, I., Landau, E., Shtutman, M., Cohen, M., Arnold, K., Feinstein, E., Geiger, B., and Bershadsky, A. (2007). p120 catenin regulates lamellipodial dynamics and cell adhesion in cooperation with cortactin. Proc. Natl. Acad. Sci. USA *104*, 10882–10887.

Brasch, J., Harrison, O.J., Ahlsen, G., Carnally, S.M., Henderson, R.M., Honig, B., and Shapiro, L. (2011). Structure and binding mechanism of vascular endothelial cadherin: a divergent classical cadherin. J. Mol. Biol. *408*, 57–73.

Bryson-Richardson, R.J. and Currie, P.D. (2008). The genetics of vertebrate myogenesis. Nat. Rev. Genet. *9*, 632–646.

Buckingham, M. (2001). Skeletal muscle formation in vertebrates. Curr. Opin. Genet. Dev. *11*, 440–448.

Buckingham, M. (2006). Myogenic progenitor cells and skeletal myogenesis in vertebrates. Curr. Opin. Genet. Dev. *16*, 525–532.

Burridge, K. and Wennerberg, K. (2004). Rho and Rac take center stage. Cell *116*, 167–179.

Burute, M. and Thery, M. (2012). Spatial segregation between cell-cell and cell-matrix adhesions. Curr. Opin. Cell Biol. *24*, 628–636.

Bustos, R.I., Forget, M.A., Settleman, J.E., and Hansen, S.H. (2008). Coordination of Rho and Rac GTPase function via p190B RhoGAP. Curr. Biol. *18*, 1606–1611.

Cadot, B., Gache, V., Vasyutina, E., Falcone, S., Birchmeier, C., and Gomes, E.R. (2012). Nuclear movement during myotube formation is microtubule and dynein dependent and is regulated by Cdc42, Par6 and Par3. EMBO Rep. *13*, 741–749.

Carmona-Fontaine, C., Matthews, H.K., Kuriyama, S., Moreno, M., Dunn, G.A., Parsons, M., Stern, C.D., and Mayor, R. (2008). Contact inhibition of locomotion *in vivo* controls neural crest directional migration. Nature *456*, 957–961.

Charrasse, S., Comunale, F., Fortier, M., Portales-Casamar, E., Debant, A., and Gauthier-Rouviere, C. (2007). M-cadherin activates Rac1 GTPase through the Rho-GEF trio during myoblast fusion. Mol. Biol. Cell *18*, 1734–1743.

Charrasse, S., Comunale, F., Grumbach, Y., Poulat, F., Blangy, A., and Gauthier-Rouviere, C. (2006). RhoA GTPase regulates M-cadherin activity and myoblast fusion. Mol. Biol. Cell *17*, 749–759.

Charrasse, S., Meriane, M., Comunale, F., Blangy, A., and Gauthier-Rouviere, C. (2002). N-cadherin-dependent cell-cell contact regulates Rho GTPases and beta-catenin localization in mouse C2C12 myoblasts. J. Cell Biol. *158*, 953–965.

Chen, X.L., Nam, J.O., Jean, C., Lawson, C., Walsh, C.T., Goka, E., Lim, S.T., Tomar, A., Tancioni, I., Uryu, S., *et al.* (2012). VEGF-induced vascular permeability is mediated by FAK. Dev. Cell *22*, 146–157.

Chircop, M. (2014). Rho GTPases as regulators of mitosis and cytokinesis in mammalian cells. Small GTPases *5*, e29770.

Choi, W., Acharya, B.R., Peyret, G., Fardin, M.A., Mege, R.M., Ladoux, B., Yap, A.S., Fanning, A.S., and Peifer, M. (2016). Remodeling the zonula adherens in response to tension and the role of afadin in this response. J. Cell Biol. *213*, 243–260.

Christian, L., Bahudhanapati, H., and Wei, S. (2013). Extracellular metalloproteinases in neural crest development and craniofacial morphogenesis. Crit. Rev. Biochem. Mol. Biol. *48*, 544–560.

Citi, S., Guerrera, D., Spadaro, D., and Shah, J. (2014). Epithelial junctions and Rho family GTPases: the zonular signalosome. Small GTPases *5*, 1–15.

D'Souza-Schorey, C. and Chavrier, P. (2006). ARF proteins: roles in membrane traffic and beyond. Nat. Rev. Mol. Cell Biol. *7*, 347–358.

Diz-Munoz, A., Thurley, K., Chintamen, S., Altschuler, S.J., Wu, L.F., Fletcher, D.A., and Weiner, O.D. (2016). Membrane tension acts through PLD2 and mTORC2 to limit actin network assembly during neutrophil migration. PLoS Biol. *14*, e1002474.

Doherty, J.T., Lenhart, K.C., Cameron, M.V., Mack, C.P., Conlon, F.L., and Taylor, J.M. (2011). Skeletal muscle differentiation and fusion are regulated by the BAR-containing Rho-GTPase-activating protein (Rho-GAP), GRAF1. J. Biol. Chem. *286*, 25903–25921.

Donaldson, J.G. (2008). Arfs and membrane lipids: sensing, generating and responding to membrane curvature. Biochem. J. *414*, e1–e2.

Duan, R. and Gallagher, P.J. (2009). Dependence of myoblast fusion on a cortical actin wall and nonmuscle myosin IIA. Dev. Biol. *325*, 374–385.

Dubash, A.D., Menold, M.M., Samson, T., Boulter, E., Garcia-Mata, R., Doughman, R., and Burridge, K. (2009). Chapter 1. Focal adhesions: new angles on an old structure. Int. Rev. Cell Mol. Biol. *277*, 1–65.

Fort, P. and Theveneau, E. (2014). PleiotRHOpic: Rho pathways are essential for all stages of neural crest development. Small GTPases *5*, e27975.

Fortier, M., Comunale, F., Kucharczak, J., Blangy, A., Charrasse, S., and Gauthier-Rouviere, C. (2008). RhoE controls myoblast alignment prior fusion through RhoA and ROCK. Cell Death Differ. *15*, 1221–1231.

Fritz, R.D., Menshykau, D., Martin, K., Reimann, A., Pontelli, V., and Pertz, O. (2015). SrGAP2-dependent integration of membrane geometry and slit-robo-repulsive cues regulates fibroblast contact inhibition of locomotion. Dev. Cell *35*, 78–92.

Fukuyama, T., Ogita, H., Kawakatsu, T., Inagaki, M., and Takai, Y. (2006). Activation of Rac by cadherin through the c-Src-Rap1-phosphatidylinositol 3-kinase-Vav2 pathway. Oncogene *25*, 8–19.

Giepmans, B.N. and van Ijzendoorn, S.C. (2009). Epithelial cell-cell junctions and plasma membrane domains. Biochim. Biophys. Acta *1788*, 820–831.

Gopal, S., Multhaupt, H.A., Pocock, R., and Couchman, J.R. (2016). Cell-extracellular matrix and cell-cell adhesion are linked by syndecan-4. Matrix Biol. (journal of the International Society for Matrix Biology) *60–61*, 57–69.

Grosheva, I., Shtutman, M., Elbaum, M., and Bershadsky, A.D. (2001). p120 catenin affects cell motility via modulation of activity of Rho-family GTPases: a link between cell-cell contact formation and regulation of cell locomotion. J. Cell Sci. *114*, 695–707.

Hamoud, N., Tran, V., Croteau, L.P., Kania, A., and Cote, J.F. (2014). G-protein coupled receptor BAI3 promotes myoblast fusion in vertebrates. Proc. Natl. Acad. Sci. USA *111*, 3745–3750.

Helbling, P.M., Saulnier, D.M., Robinson, V., Christiansen, J.H., Wilkinson, D.G., and Brandli, A.W. (1999). Comparative analysis of embryonic gene expression defines potential interaction sites for *Xenopus* EphB4 receptors with ephrin-B ligands. Dev. Dyn. (an official publication of the American Association of Anatomists) *216*, 361–373.

Hinde, E., Digman, M.A., Hahn, K.M., and Gratton, E. (2013). Millisecond spatiotemporal dynamics of FRET biosensors by the pair correlation function and the phasor approach to FLIM. Proc. Natl. Acad. Sci. USA *110*, 135–140.

Hindi, S.M., Tajrishi, M.M., and Kumar, A. (2013). Signaling mechanisms in mammalian myoblast fusion. Sci. Signal. *6*, re2.

Hochreiter-Hufford, A.E., Lee, C.S., Kinchen, J.M., Sokolowski, J.D., Arandjelovic, S., Call, J.A., Klibanov, A.L., Yan, Z., Mandell, J.W., and Ravichandran, K.S. (2013). Phosphatidylserine receptor BAI1 and apoptotic cells as new promoters of myoblast fusion. Nature *497*, 263–267.

Hocking, J.C., Hehr, C.L., Bertolesi, G.E., Wu, J.Y., and McFarlane, S. (2010). Distinct roles for Robo2 in the regulation of axon and dendrite growth by retinal ganglion cells. Mech. Dev. *127*, 36–48.

Hoon, J.L., Tan, M.H., and Koh, C.G. (2016). The regulation of cellular responses to mechanical cues by Rho GTPases. Cells *5*, 17.

Hordijk, P.L., ten Klooster, J.P., van der Kammen, R.A., Michiels, F., Oomen, L.C., and Collard, J.G. (1997). Inhibition of invasion of epithelial cells by Tiam1-Rac signaling. Science *278*, 1464–1466.

Houk, A.R., Jilkine, A., Mejean, C.O., Boltyanskiy, R., Dufresne, E.R., Angenent, S.B., Altschuler, S.J., Wu, L.F., and Weiner, O.D. (2012). Membrane tension maintains cell polarity by confining signals to the leading edge during neutrophil migration. Cell *148*, 175–188.

Huang, C., Kratzer, M.C., Wedlich, D., and Kashef, J. (2016). E-cadherin is required for cranial neural crest migration in *Xenopus laevis*. Dev. Biol. *411*, 159–171.

Ishiyama, N., Tanaka, N., Abe, K., Yang, Y.J., Abbas, Y.M., Umitsu, M., Nagar, B., Bueler, S.A., Rubinstein, J.L., Takeichi, M., *et al.* (2013). An autoinhibited structure of alpha-catenin and its implications for vinculin recruitment to adherens junctions. J. Biol. Chem. *288*, 15913–15925.

Izumi, G., Sakisaka, T., Baba, T., Tanaka, S., Morimoto, K., and Takai, Y. (2004). Endocytosis of E-cadherin regulated by Rac and Cdc42 small G proteins through IQGAP1 and actin filaments. J. Cell Biol. *166*, 237–248.

Kadir, S., Astin, J.W., Tahtamouni, L., Martin, P., and Nobes, C.D. (2011). Microtubule remodelling is required for the front-rear polarity switch during contact inhibition of locomotion. J. Cell Sci. *124*, 2642–2653.

Kanchanawong, P., Shtengel, G., Pasapera, A.M., Ramko, E.B., Davidson, M.W., Hess, H.F., and Waterman, C.M. (2010). Nanoscale architecture of integrin-based cell adhesions. Nature *468*, 580–584.

Kashef, J., Kohler, A., Kuriyama, S., Alfandari, D., Mayor, R., and Wedlich, D. (2009). Cadherin-11 regulates protrusive activity in *Xenopus* cranial neural crest cells upstream of Trio and the small GTPases. Genes Dev. *23*, 1393–1398.

Kim, J.H., Jin, P., Duan, R., and Chen, E.H. (2015). Mechanisms of myoblast fusion during muscle development. Curr. Opin. Genet. Dev. *32*, 162–170.

Kitt, K.N. and Nelson, W.J. (2011). Rapid suppression of activated Rac1 by cadherins and nectins during de novo cell-cell adhesion. PLoS One *6*, e17841.

Kiuchi, T., Nagai, T., Ohashi, K., and Mizuno, K. (2011). Measurements of spatiotemporal changes in G-actin concentration reveal its effect on stimulus-induced actin assembly and lamellipodium extension. J. Cell Biol. *193*, 365–380.

Kooistra, M.R., Dube, N., and Bos, J.L. (2007). Rap1: a key regulator in cell-cell junction formation. J. Cell Sci. *120*, 17–22.

Kourtidis, A., Ngok, S.P., and Anastasiadis, P.Z. (2013). p120 catenin: an essential regulator of cadherin stability, adhesion-induced signaling, and cancer progression. Prog. Mol. Biol. Transl. Sci. *116*, 409–432.

Kovacs, E.M., Ali, R.G., McCormack, A.J., and Yap, A.S. (2002). E-cadherin homophilic ligation directly signals through Rac and phosphatidylinositol 3-kinase to regulate adhesive contacts. J. Biol. Chem. *277*, 6708–6718.

Kuriyama, S., Theveneau, E., Benedetto, A., Parsons, M., Tanaka, M., Charras, G., Kabla, A., and Mayor, R. (2014). *In vivo* collective cell migration requires an LPAR2-dependent increase in tissue fluidity. J. Cell Biol. *206*, 113–127.

Lakhani, V.V., Hinde, E., Gratton, E., and Elston, T.C. (2015). Spatio-temporal regulation of Rac1 mobility by actin islands. PLoS One *10*, e0143753.

Laurin, M., Fradet, N., Blangy, A., Hall, A., Vuori, K., and Cote, J.F. (2008). The atypical Rac activator Dock180 (Dock1) regulates myoblast fusion *in vivo*. Proc. Natl. Acad. Sci. USA *105*, 15446–15451.

Leckband, D.E. and de Rooij, J. (2014). Cadherin adhesion and mechanotransduction. Annu. Rev. Cell Dev. Biol. *30*, 291–315.

Lenhart, K.C., Becerer, A.L., Li, J., Xiao, X., McNally, E.M., Mack, C.P., and Taylor, J.M. (2014). GRAF1 promotes ferlin-dependent myoblast fusion. Dev. Biol. *393*, 298–311.

Liu, Z., Tan, J.L., Cohen, D.M., Yang, M.T., Sniadecki, N.J., Ruiz, S.A., Nelson, C.M., and Chen, C.S. (2010). Mechanical tugging force regulates the size of cell-cell junctions. Proc. Natl. Acad. Sci. USA *107*, 9944–9949.

Mack, N.A. and Georgiou, M. (2014). The interdependence of the Rho GTPases and apicobasal cell polarity. Small GTPases 5, 10.

Malliri, A., van Es, S., Huveneers, S., and Collard, J.G. (2004). The Rac exchange factor Tiam1 is required for the establishment and maintenance of cadherin-based adhesions. J. Biol. Chem. 279, 30092–30098.

Matthews, H.K., Marchant, L., Carmona-Fontaine, C., Kuriyama, S., Larrain, J., Holt, M.R., Parsons, M., and Mayor, R. (2008). Directional migration of neural crest cells *in vivo* is regulated by Syndecan-4/Rac1 and non-canonical Wnt signaling/RhoA. Development 135, 1771–1780.

Mayor, R. and Carmona-Fontaine, C. (2010). Keeping in touch with contact inhibition of locomotion. Trends Cell Biol. 20, 319–328.

McLennan, I.S. and Koishi, K. (2002). The transforming growth factor-betas: multifaceted regulators of the development and maintenance of skeletal muscles, motoneurons and Schwann cells. Int. J. Dev. Biol. 46, 559–567.

Meriane, M., Roux, P., Primig, M., Fort, P., and Gauthier-Rouviere, C. (2000). Critical activities of Rac1 and Cdc42Hs in skeletal myogenesis: antagonistic effects of JNK and p38 pathways. Mol. Biol. Cell 11, 2513–2528.

Mohr, M., Zaenker, K.S., and Dittmar, T. (2015). Fusion in cancer: an explanatory model for aneuploidy, metastasis formation, and drug resistance. Method. Mol. Biol. 1313, 21–40.

Moore, C.A., Parkin, C.A., Bidet, Y., and Ingham, P.W. (2007). A role for the myoblast city homologues Dock1 and Dock5 and the adaptor proteins Crk and Crk-like in zebrafish myoblast fusion. Development 134, 3145–3153.

Moore, R., Theveneau, E., Pozzi, S., Alexandre, P., Richardson, J., Merks, A., Parsons, M., Kashef, J., Linker, C., and Mayor, R. (2013). Par3 controls neural crest migration by promoting microtubule catastrophe during contact inhibition of locomotion. Development 140, 4763–4775.

Mui, K.L., Chen, C.S., and Assoian, R.K. (2016). The mechanical regulation of integrin-cadherin crosstalk organizes cells, signaling and forces. J. Cell Sci. 129, 1093–1100.

Nanes, B.A., Chiasson-MacKenzie, C., Lowery, A.M., Ishiyama, N., Faundez, V., Ikura, M., Vincent, P.A., and Kowalczyk, A.P. (2012). p120-catenin binding masks an endocytic signal conserved in classical cadherins. J. Cell Biol. 199, 365–380.

Nelson, W.J. (2008). Regulation of cell-cell adhesion by the cadherin-catenin complex. Biochem. Soc. Trans. 36, 149–155.

Neuner, R., Cousin, H., McCusker, C., Coyne, M., and Alfandari, D. (2009). *Xenopus* ADAM19 is involved in neural, neural crest and muscle development. Mech. Dev. 126, 240–255.

Nishiyama, T., Kii, I., and Kudo, A. (2004). Inactivation of Rho/ROCK signaling is crucial for the nuclear accumulation of FKHR and myoblast fusion. J. Biol. Chem. 279, 47311–47319.

Nola, S., Daigaku, R., Smolarczyk, K., Carstens, M., Martin-Martin, B., Longmore, G., Bailly, M., and Braga, V.M. (2011). Ajuba is required for Rac activation and maintenance of E-cadherin adhesion. J. Cell Biol. 195, 855–871.

Noren, N.K., Arthur, W.T., and Burridge, K. (2003). Cadherin engagement inhibits RhoA via p190RhoGAP. J. Biol. Chem. *278*, 13615–13618.

Noren, N.K., Niessen, C.M., Gumbiner, B.M., and Burridge, K. (2001). Cadherin engagement regulates Rho family GTPases. J. Biol. Chem. *276*, 33305–33308.

Nowak, S.J., Nahirney, P.C., Hadjantonakis, A.K., and Baylies, M.K. (2009). Nap1-mediated actin remodeling is essential for mammalian myoblast fusion. J. Cell Sc. *122*, 3282–3293.

O'Brien, S.P., Seipel, K., Medley, Q.G., Bronson, R., Segal, R., and Streuli, M. (2000). Skeletal muscle deformity and neuronal disorder in Trio exchange factor-deficient mouse embryos. Proc. Natl. Acad. Sci. USA *97*, 12074–12078.

Omelchenko, T. and Hall, A. (2012). Myosin-IXA regulates collective epithelial cell migration by targeting RhoGAP activity to cell-cell junctions. Curr. Biology. *22*, 278–288.

Ouyang, M., Lu, S., Kim, T., Chen, C.E., Seong, J., Leckband, D.E., Wang, F., Reynolds, A.B., Schwartz, M.A., and Wang, Y. (2013). N-cadherin regulates spatially polarized signals through distinct p120ctn and beta-catenin-dependent signalling pathways. Nat. Commun. *4*, 1589.

Pajcini, K.V., Pomerantz, J.H., Alkan, O., Doyonnas, R., and Blau, H.M. (2008). Myoblasts and macrophages share molecular components that contribute to cell-cell fusion. J. Cell Biol. *180*, 1005–1019.

Patel, S.D., Ciatto, C., Chen, C.P., Bahna, F., Rajebhosale, M., Arkus, N., Schieren, I., Jessell, T.M., Honig, B., Price, S.R., *et al.* (2006). Type II cadherin ectodomain structures: implications for classical cadherin specificity. Cell *124*, 1255–1268.

Peckham, M. (2008). Engineering a multi-nucleated myotube, the role of the actin cytoskeleton. J. Microsc. *231*, 486–493.

Perez, T.D., Tamada, M., Sheetz, M.P., and Nelson, W.J. (2008). Immediate-early signaling induced by E-cadherin engagement and adhesion. Journal Biol. Chem. *283*, 5014–5022.

Ratheesh, A., Gomez, G.A., Priya, R., Verma, S., Kovacs, E.M., Jiang, K., Brown, N.H., Akhmanova, A., Stehbens, S.J., and Yap, A.S. (2012). Centralspindlin and alpha-catenin regulate Rho signalling at the epithelial zonula adherens. Nat. Cell Biol. *14*, 818–828.

Ratheesh, A., Priya, R., and Yap, A.S. (2013). Coordinating Rho and Rac: the regulation of Rho GTPase signaling and cadherin junctions. Prog. Mol. Biol. Transl. Sci. *116*, 49–68.

Richardson, B.E., Nowak, S.J., and Baylies, M.K. (2008). Myoblast fusion in fly and vertebrates: new genes, new processes and new perspectives. Traffic *9*, 1050–1059.

Rochlin, K., Yu, S., Roy, S., and Baylies, M.K. (2010). Myoblast fusion: when it takes more to make one. Dev. Biol. *341*, 66–83.

Sadok, A. and Marshall, C.J. (2014). Rho GTPases: masters of cell migration. Small GTPases *5*, e29710.

Sander, E.E., ten Klooster, J.P., van Delft, S., van der Kammen, R.A., and Collard, J.G. (1999). Rac downregulates Rho activity: reciprocal balance between both GTPases determines cellular morphology and migratory behavior. J. Cell Biol. *147*, 1009–1022.

Scarpa, E., Szabo, A., Bibonne, A., Theveneau, E., Parsons, M., and Mayor, R. (2015). Cadherin switch during EMT in neural crest cells leads to contact inhibition of locomotion via repolarization of forces. Dev. Cell *34*, 421–434.

Schlessinger, K., Hall, A., and Tolwinski, N. (2009). Wnt signaling pathways meet Rho GTPases. Genes Dev. *23*, 265–277.

Sebe, A., Erdei, Z., Varga, K., Bodor, C., Mucsi, I., and Rosivall, L. (2010). Cdc42 regulates myocardin-related transcription factor nuclear shuttling and alpha-smooth muscle actin promoter activity during renal tubular epithelial-mesenchymal transition. Nephron Exp. Nephrol. *114*, e117–e125.

Shikanai, M., Nakajima, K., and Kawauchi, T. (2011). N-cadherin regulates radial glial fiber-dependent migration of cortical locomoting neurons. Commun. Integr. Biol. *4*, 326–330.

Shoval, I., Ludwig, A., and Kalcheim, C. (2007). Antagonistic roles of full-length N-cadherin and its soluble BMP cleavage product in neural crest delamination. Development *134*, 491–501.

Smith, A., Robinson, V., Patel, K., and Wilkinson, D.G. (1997). The EphA4 and EphB1 receptor tyrosine kinases and ephrin-B2 ligand regulate targeted migration of branchial neural crest cells. Curr. Biol. *7*, 561–570.

Srinivas, B.P., Woo, J., Leong, W.Y., and Roy, S. (2007). A conserved molecular pathway mediates myoblast fusion in insects and vertebrates. Nat. Genet. *39*, 781–786.

Stehbens, S.J., Paterson, A.D., Crampton, M.S., Shewan, A.M., Ferguson, C., Akhmanova, A., Parton, R.G., and Yap, A.S. (2006). Dynamic microtubules regulate the local concentration of E-cadherin at cell-cell contacts. J. Cell Sci. *119*, 1801–1811.

Stramer, B., and Mayor, R. (2016). Mechanisms and *in vivo* functions of contact inhibition of locomotion. Nat. Rev. Mol. Cell Biol. doi: 10.1038/nrm.2016.118.

Swailes, N.T., Colegrave, M., Knight, P.J., and Peckham, M. (2006). Non-muscle myosins 2A and 2B drive changes in cell morphology that occur as myoblasts align and fuse. J. Cell Sci. *119*, 3561–3570.

Swailes, N.T., Knight, P.J., and Peckham, M. (2004). Actin filament organization in aligned prefusion myoblasts. J. Anat. *205*, 381–391.

Taneyhill, L.A. and Schiffmacher, A.T. (2017). Should I stay or should I go? Cadherin function and regulation in the neural crest. Genesis. doi: 10.1002/dvg.23028.

Terry, S.J., Zihni, C., Elbediwy, A., Vitiello, E., Leefa Chong San, I.V., Balda, M.S., and Matter, K. (2011). Spatially restricted activation of RhoA signalling at epithelial junctions by p114RhoGEF drives junction formation and morphogenesis. Nat. Cell Biol. *13*, 159–166.

Theveneau, E., Marchant, L., Kuriyama, S., Gull, M., Moepps, B., Parsons, M., and Mayor, R. (2010). Collective chemotaxis requires contact-dependent cell polarity. Dev. Cell *19*, 39–53.

Theveneau, E. and Mayor, R. (2012). Neural crest delamination and migration: from epithelium-to-mesenchyme transition to collective cell migration. Dev. Biology. *366*, 34–54.

Theveneau, E., Steventon, B., Scarpa, E., Garcia, S., Trepat, X., Streit, A., and Mayor, R. (2013). Chase-and-run between adjacent cell populations promotes directional collective migration. Nat. Cell Biol. *15*, 763–772.

Thorsteinsdottir, S., Deries, M., Cachaco, A.S., and Bajanca, F. (2011). The extracellular matrix dimension of skeletal muscle development. Dev. Biol. *354*, 191–207.

Timmerman, I., Heemskerk, N., Kroon, J., Schaefer, A., van Rijssel, J., Hoogenboezem, M., van Unen, J., Goedhart, J., Gadella, T.W., Jr., Yin, T., et al. (2015). A local VE-cadherin and Trio-based signaling complex stabilizes endothelial junctions through Rac1. J. Cell Sci. *128*, 3041–3054.

Trepat, X. and Fredberg, J.J. (2011). Plithotaxis and emergent dynamics in collective cellular migration. Trends Cell Biol. *21*, 638–646.

Tseng, Q., Duchemin-Pelletier, E., Deshiere, A., Balland, M., Guillou, H., Filhol, O., and Thery, M. (2012). Spatial organization of the extracellular matrix regulates cell-cell junction positioning. Proc. Natl. Acad. Sci. USA *109*, 1506–1511.

Tsujita, K., Takenawa, T., and Itoh, T. (2015). Feedback regulation between plasma membrane tension and membrane-bending proteins organizes cell polarity during leading edge formation. Nat. Cell Biol. *17*, 749–758.

Tu, C.L. and You, M. (2014). Obligatory roles of filamin A in E-cadherin-mediated cell-cell adhesion in epidermal keratinocytes. J. Dermatol. Sci. *73*, 142–151.

Vasyutina, E., Martarelli, B., Brakebusch, C., Wende, H., and Birchmeier, C. (2009). The small G-proteins Rac1 and Cdc42 are essential for myoblast fusion in the mouse. Proc. Natl. Acad. Sci. USA *106*, 8935–8940.

Weber, G.F., Bjerke, M.A., and DeSimone, D.W. (2012). A mechanoresponsive cadherin-keratin complex directs polarized protrusive behavior and collective cell migration. Dev. Cell *22*, 104–115.

Wei, L., Zhou, W., Croissant, J.D., Johansen, F.E., Prywes, R., Balasubramanyam, A., and Schwartz, R.J. (1998). RhoA signaling via serum response factor plays an obligatory role in myogenic differentiation. J. Biol. Chem. *273*, 30287–30294.

Wildenberg, G.A., Dohn, M.R., Carnahan, R.H., Davis, M.A., Lobdell, N.A., Settleman, J., and Reynolds, A.B. (2006). p120-catenin and p190RhoGAP regulate cell-cell adhesion by coordinating antagonism between Rac and Rho. Cell *127*, 1027–1039.

Xu, C., Funahashi, Y., Watanabe, T., Takano, T., Nakamuta, S., Namba, T., and Kaibuchi, K. (2015). Radial glial cell-neuron interaction directs axon formation at the opposite side of the neuron from the contact site. J. Neurosci. (the official journal of the Society for Neuroscience) *35*, 14517–14532.

Yamada, S. and Nelson, W.J. (2007). Localized zones of Rho and Rac activities drive initiation and expansion of epithelial cell-cell adhesion. J. Cell Biol. *178*, 517–527.

Yano, H., Mazaki, Y., Kurokawa, K., Hanks, S.K., Matsuda, M., and Sabe, H. (2004). Roles played by a subset of integrin signaling molecules in cadherin-based cell-cell adhesion. J. Cell Biol. *166*, 283–295.

Zebda, N., Tian, Y., Tian, X., Gawlak, G., Higginbotham, K., Reynolds, A.B., Birukova, A.A., and Birukov, K.G. (2013). Interaction of p190RhoGAP with C-terminal domain of p120-catenin modulates endothelial cytoskeleton and permeability. The J. Biol. Chem. *288*, 18290–18299.

Rho signaling in mechanotransduction

5

Elizabeth Monaghan-Benson* and Christophe Guilluy[†,‡]

*Department of Cell Biology and Physiology, University of North Carolina at Chapel Hill, Chapel Hill, North Carolina

[†] *Institute for Advanced Biosciences, Inserm U1209, CNRS UMR 5309, Université Grenoble Alpes, Grenoble 38000, France*

[‡] *christophe.guilluy@inserm.fr*

Keywords: Mechanotransduction, adhesion, integrin, RhoA, Rac, cytoskeleton.

5.1. Introduction

The ability of cells to respond to mechanical stress is fundamental to many biological processes from development to disease. Mechanical forces are transmitted and transduced into specific biochemical signals that ultimately affect cellular function. Over the last three decades, studies have demonstrated that mechanotransduction mechanisms modulate many aspects of cell growth and behavior, ranging from the assembly of cytoskeletal structures to the regulation of gene expression.

Since the early work of Anne Ridley and Alan Hall, RhoA and Rac1, two members of the Rho protein family have been implicated in mechanotransduction. Rho GTPases are a subgroup of the Ras superfamily that act as molecular switches cycling between an active GTP-bound state and an inactive GDP-bound state (Jaffe and Hall, 2005). Once activated, Rho proteins interact with downstream effectors to trigger specific signaling pathways and elicit cellular responses.

The seminal work of Hall and Ridley drew attention to Rho proteins and revealed that they participate in the assembly of stress fibers (SF) and focal adhesions (Ridley and Hall, 1992), two subcellular structures that play a central role in mechanotransduction by generating and transmitting mechanical tension. Numerous studies have now shown that a wide range of mechanical forces regulate the activity of Rho proteins, which in turn signal to downstream effectors and mediate the cellular response to mechanical stress.

After a short overview of mechanical forces in biology, we will describe how the activities of RhoA and Rac1 are regulated downstream from distinct mechanosensitive structures by GEFs and GAPs. Focusing on RhoA and Rac1, we will discuss how their effectors mediate the cellular response to force. Additionally, we will discuss how these Rho-mediated mechanotransduction mechanisms participate in the pathogenesis of diseases.

5.2. Mechanical stress in biology

5.2.1. *Mechanical force, stress and matrix compliance*

Force is classically defined as an interaction that causes an object with mass to change its velocity. Forces are characterized by a vector with a direction and a magnitude, whereas stress represents a force per unit of area. Tension and compression are applied perpendicularly to the cell and lead to its expansion and compaction respectively, whereas shear stress is applied parallel to the surface of the cell. Cells are continuously subjected to a wide variety of mechanical forces that can be externally applied or internally generated by the cell's own actomyosin contractile system. External forces depend upon the cellular microenvironment (Fig. 1); for example, chondrocytes experience compressive forces from body weight and muscle tension, while lung epithelial cells are exposed to continuous cyclic tension from the expansion of the lung during breathing.

In addition to externally applied forces, cells generate myosin-dependent contractility, which results in tension applied at sites of cell–matrix and cell–cell adhesion (Burridge and Wittchen, 2013). Harris and colleagues first reported and quantified these forces by observing the wrinkles formed when cells adhere to thin films of silicone *rubber* (Harris *et al.*, 1980). Continued studies have resulted in the development of tools to measure tension on single adhesion molecules or individual adhesion complexes, including traction force microscopy (Balaban *et al.*, 2001) and fluorescence resonance energy transfer (FRET) tension sensors (Grashoff *et al.*, 2010). Cell-generated contractility is also influenced by the stiffness of the extracellular matrix (ECM). The stiffness, or rigidity, of the ECM indicates its

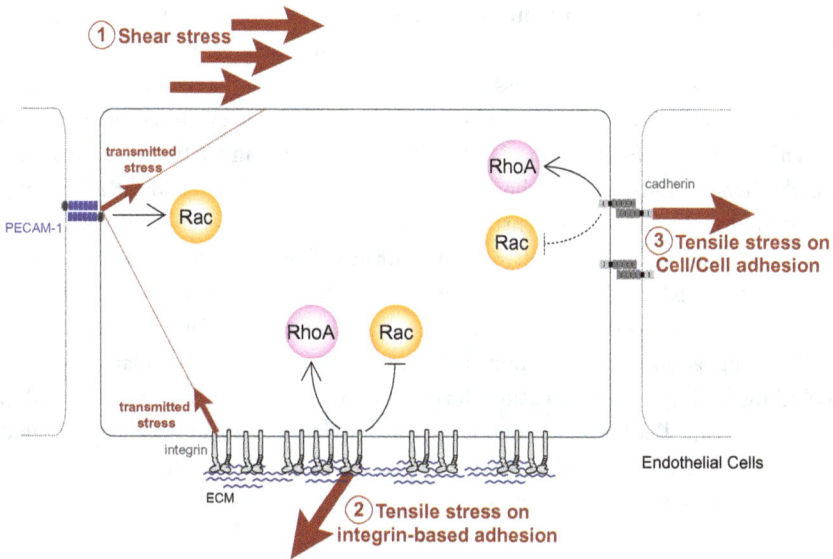

Figure 1. Mechanical forces regulate Rho protein activity. Cells are subjected to a wide range of external forces depending on their microenvironment; as an example, endothelial cells experience hemodynamic forces in the form of shear and tensile stress due to the pulsatile nature of blood motion. (1) Shear stress transmission to PECAM-1 activates Rac1 via Src-dependent vav2 activation (Liu *et al.*, 2013). Application of tension to PECAM-1 indirectly impacts RhoA activity after stress propagation to integrin-based adhesions (Collins *et al.*, 2012). Application of mechanical tension to the adhesion complex (2) activates RhoA (Guilluy *et al.*, 2011; Matthews *et al.*, 2006; Zhao *et al.*, 2007). In contrast to its effect on RhoA activity, mechanical tension applied to integrin-based adhesion decreases Rac1 activity (Katsumi *et al.*, 2002). (3) Tensional forces applied to cadherin-based adhesions increase RhoA activity (Abiko *et al.*, 2015; Liu *et al.*, 2007; Marjoram *et al.*, 2015), whereas it was shown to either increase (Cai *et al.*, 2014; Liu *et al.*, 2007) or decrease Rac1 activity (Marjoram *et al.*, 2015).

resistance to deformation in response to the application of force. The inverse of stiffness is compliance. Cell-generated contractility plays a critical role in substrate elasticity sensing, as cells continuously adjust to the stiffness of their physical environment by generating tension within the cytoskeleton.

The Rho GTPases are key regulators of myosin II–generated tension. Downstream of RhoA, Rho kinase (ROCK) 1 and 2 promote myosin light chain (MLC) phosphorylation, directly and indirectly by inhibiting MLC phosphatase. The Rac1 and Cdc42 effector PAK can have opposite effects on MLC phosphorylation depending on the cell type analyzed, whereas MRCK, another Cdc42 effector, phosphorylates MLC (Vicente-Manzanares *et al.*, 2009). MLC phosphorylation enhances the assembly of myosin II into filaments and promotes its ATPase activity thereby increasing the contractile force exhibited by myosin II on actin filaments. Furthermore, ROCK phosphorylates and activates LIM kinase which then inhibits cofilin, the actin severing protein (Maekawa *et al.*, 1999). Additionally, RhoA promotes further actin filament formation through its effector mDia, an actin nucleating protein in the formin family (Watanabe *et al.*, 1999). Consequently, RhoA signaling is largely responsible for much of the intracellular force generation within cells (Burridge and Wittchen, 2013). Forces are also generated during actin polymerization and contribute to diverse cellular processes, including leading edge protrusion during migration (Mogilner and Oster, 2003). Both RhoA and Rac1 stimulate actin nucleation via the regulation of actin-related protein 2/3 (Arp2/3) and formins, respectively, thereby promoting the force generation associated with actin polymerization.

5.2.2. *Mechanotransmission and mechanosensing*

Mechanotransduction is mediated by molecular mechanisms that transmit and transduce mechanical signals into specific biochemical pathways (Hoffman *et al.*, 2011). Pioneering work showed that mechanical stress transmission can occur at relatively long distances within the cell and that forces applied to surface adhesions can be propagated to inner organelles (Wang *et al.*, 1993). The level of isometric tension, or prestress, within the cytoskeleton greatly impacts mechanical stress transmission. Depending on the extent of mechanical stress propagation, distinct mechanosensitive elements will be subjected to tension, and this will determine and orientate the cellular response.

Mechanosensing hinges on proteins whose conformations are affected by mechanical tension, leading to changes in activity, subcellular localization, or interacting partners. Mechanosensors are typically associated with force-bearing subcellular structures, such as the cell surface lipid membrane which anchors mechanosensitive ion channels (Volkers *et al.*, 2015). As a physical connection between the extracellular matrix and the cytoskeleton, integrin-based adhesion experiences tension emanating from the environment and from actomyosin filaments. Using beads coated with integrin ligands, studies have shown that the application of force to integrins induces cytoskeletal rearrangements of the adhesion complex, resulting in increased local stiffness, also known as reinforcement (Choquet *et al.*, 1997; Wang *et al.*, 1993). This mechanical response is associated with changes in adhesion composition in the process of adhesion maturation

(Kuo *et al.*, 2011). Tension-induced adhesion reinforcement is mediated by adhesion components that act as mechanosensors. The quintessential example of adhesion mechanosensor is talin. Single molecule force manipulation demonstrated that application of tension produces an extension of the rod domain and exposes new vinculin binding sites, leading to vinculin recruitment and adhesion growth. In addition to talin, many other adhesion components behave as mechanosensors, including integrins, integrin ligands, and scaffolding proteins like p130Cas and confer mechanosensitivity to the whole adhesion complex (Hoffman *et al.*, 2011). Cadherin–catenin complexes at sites of cell–cell adhesions are also mechanosensitive structures, whose organization is regulated by mechanical tension (Ladoux *et al.*, 2010; Liu *et al.*, 2010). Within this complex, α-catenin operates a conformational switch that participates in vinculin recruitment and adhesion reinforcement in response to tension (Kim *et al.*, 2015). PECAM-1 also acts as a mechanosensor at endothelial cell–cell junctions, where it triggers the endothelial response to shear stress (Tzima *et al.*, 2005) (Fig. 1). While other subcellular structures are known to be mechanosensitive, including caveolae, the lamina, stress fibers, and the primary cilia, the identities of the specific mechanosensors involved in these structures remain unknown.

5.3. Regulation of Rho protein activity in response to mechanical tension

The mechanisms regulating Rho protein activity were initially characterized downstream of soluble factors, such as growth factors and hormones, but the development of experimental systems that stimulate cells with mechanical stress has offered insights into force-induced Rho protein regulation. The activity of Rho GTPases is regulated by three classes of proteins: guanine nucleotide exchange factors (GEFs), GTPase activating proteins (GAPs) and guanine nucleotide dissociation inhibitors (GDIs) (see Chapter 3 by Amin and Ahmadian). GEFs activate Rho GTPases by promoting the exchange of GDP for GTP, whereas GAPs stimulate the GTPase activity of Rho proteins thereby converting GTP to GDP and inactivating them (Jaffe and Hall, 2005). GDIs extract membrane-bound GTPases into the cytosol, where they are sequestered in their inactive conformation (Garcia-Mata *et al.*, 2011). Additionally, Rho proteins can undergo posttranslational modifications that eventually impact their interaction with effectors. A variety of GEFs and GAPs are regulated downstream of mechanosensitive structures and lend spatial and temporal control to regulating RhoA and Rac1 activity and initiate the mechanoresponse.

5.3.1. Regulation of Rho activity in response to tension on integrins

RhoA and Rac1 activity is regulated in a biphasic fashion during integrin-mediated adhesion. Integrin engagement and early adhesion are characterized by transient RhoA inhibition and Rac1 activation, followed by RhoA activation and Rac1 inhibition associated with an increased cell contractility (Lawson and Burridge, 2014). During this second phase, mechanical tension triggers positive feedback to RhoA and stimulates its activity promoting adhesion maturation (Geiger *et al.*, 2009; Riveline *et al.*, 2001). Application of mechanical tension to the adhesion complex, whether externally applied or cell-generated, activates RhoA (Guilluy *et al.*, 2011; Matthews *et al.*, 2006; Zhao *et al.*, 2007) (Fig. 1). Tensional forces applied to integrins induce the recruitment of the RhoA GEFs, GEF-H1/ARHGEF2, and leukemia-associated Rho guanine nucleotide exchange factor (LARG/ARHGEF12) to adhesion complexes, where both mediate force-induced RhoA activation (Guilluy *et al.*, 2011). The tyrosine kinase Fyn activates LARG, while the FAK/Ras/ERK pathway stimulates GEF-H1 catalytic activity in response to tension (Guilluy *et al.*, 2011). Consistent with this, myosin-dependent cell contractility induces GEF-H1 enrichment at adhesion complexes supported by αv-class integrins (Kuo *et al.*, 2011; Schiller *et al.*, 2013). Additionally, GEF-H1 catalytic activity is increased in epithelial cells cultured on rigid ECM and mediates RhoA activation in response to matrix stiffness (Heck *et al.*, 2012). Cyclic stretch generates transient tension on both integrin-based adhesions and cell–cell adhesions. Numerous GEFs have been shown to contribute to RhoA activation in cells stimulated with stretch; among them are GEF-H1 (Gawlak *et al.*, 2014); LARG, VAV2 (Peng *et al.*, 2010); and SOLO/ARHGEF40 (Abiko *et al.*, 2015). However, SOLO-mediated RhoA activation in response to stretch in endothelial cells requires intact VE-cadherin adhesions, suggesting that SOLO is activated in response to tension at the site of cell–cell adhesion (Abiko *et al.*, 2015).

In contrast to its effect on RhoA activity, mechanical tension applied to integrin-based adhesion decreases Rac1 activity (Katsumi *et al.*, 2002) (Fig. 1). The Rac1 GEF betaPix/ARHGEF7 is less abundant in adhesion complexes that experience myosin-dependent tension, thus decreasing Rac1 activity in response to tension-induced adhesion maturation (Kuo *et al.*, 2011). FilGAP/ARHGAP24, a Rac-specific GAP, contributes to Rac1 inactivation in response to tensional forces (Shifrin *et al.*, 2009).

5.3.2. Regulation of Rho activity in response to tension on cell–cell adhesions

Tensional force applied to cadherin-based adhesions increase RhoA activity (Abiko *et al.*, 2015; Liu *et al.*, 2007; Marjoram *et al.*, 2015). GEFs and GAPs specific for

RhoA are localized at cell–cell junctions, although the mechanism of this tension-dependent RhoA activation is still elusive and may involve SOLO (Abiko *et al.*, 2015). Tension at sites of cell–cell adhesion was shown to either increase (Cai *et al.*, 2014; Liu *et al.*, 2007) or decrease Rac1 activity (Marjoram *et al.*, 2015) (Fig. 1). During collective cell migration, Rac1 activity is regulated in a biphasic fashion depending on the level tension applied to cadherin-based adhesion via a pathway that involves merlin (Das *et al.*, 2015).

In endothelial cells, shear stress is transmitted to cell–cell adhesion components, where it stimulates the mechanosensor PECAM-1 (Tzima *et al.*, 2005). Shear stress transmission to PECAM-1 activates Rac1 via Src-dependent VAV2 activation (Liu *et al.*, 2013). Interestingly, the GEF TIAM1 is required to direct localized activation of Rac1 in response to shear stress (Liu *et al.*, 2013). Application of tension to PECAM-1 indirectly impacts RhoA activity after stress propagation to integrin-based adhesions and stimulates GEFH1 and LARG catalytic activities (Collins *et al.*, 2012). This mechanism may contribute to the stimulatory effect of shear stress on RhoA activity (Tzima, 2006). At the surface of endothelial cells, intercellular adhesion molecule 1 (ICAM-1) and junctional adhesion protein A (JAM-A) mediate adhesion with leukocytes and facilitate transendothelial migration. Interestingly, both molecules behave as mechanosensitive structures and activate RhoA in response to tension (Lessey-Morillon *et al.*, 2014; Scott *et al.*, 2016). Whereas application of tensional forces to ICAM-1 activates RhoA via LARG (Lessey-Morillon *et al.*, 2014), the RhoA GEFs p115/ARHGEF1 and GEF-H1 are responsible for tension-dependent RhoA activation downstream of JAM-A (Scott *et al.*, 2016).

5.3.3. *Other mechanosensitive structures regulating Rho protein activity*

Caveolae are small membrane invaginations ($\approx$50–100 nm) present at the surface of many cells and are particularly abundant in cells subjected to mechanical stress, such as endothelial cells (Echarri and Del Pozo, 2015). Caveolae are mechanosensitive structures, whose curvature depends on the tension within the plasma membrane (Echarri and Del Pozo, 2015). Interestingly, caveolae are required for RhoA activation in response to stretch in cardiomyocytes and mesangial cells (Kawamura *et al.*, 2003; Peng *et al.*, 2007). Within caveolae, caveolin regulates RhoA activity through Src-dependent p190RhoGAP/ARHGAP35 phosphorylation and may mediate the tension-dependent RhoA activation downstream caveolae (Echarri and Del Pozo, 2015).

Glycocalix is a mesh of membrane-bound proteoglycans and glycoproteins that plays an important role during endothelial cell response to blood flow.

Alteration of the glycocalix affects RhoA activation in response to shear stress (Kang *et al.*, 2017); however, the molecular mechanisms regulating RhoA activity are unknown.

5.4. Rho-mediated mechanoresponse

Once activated by mechanical stress, RhoA and Rac1 interact with a variety of downstream effectors that mediate the mechanoresponse and control cellular processes at different time scale, from rapid cytoskeletal remodeling to long-term effects on transcription and differentiation.

5.4.1. *Adhesion and cytoskeletal remodeling*

RhoA-mediated SF assembly was historically observed when cells were stimulated with growth factors (Ridley and Hall, 1992), but it can occur in response to mechanical stress as well. Shear stress stimulates the formation of SFs that orient parallel to the direction of flow in endothelial cells, as observed *in vivo* (Wong *et al.*, 1983) and in cultured cells (Tzima, 2006). RhoA and its effector ROCK participate in SF formation and reorientation (Tzima, 2006) in response to shear stress applied to endothelial cells in culture. Fibroblasts exposed to cyclic stretch rearrange their SFs perpendicular to the direction of tensional force (Hayakawa *et al.*, 2001). Whereas RhoA and ROCK activations in response to cyclic stretch are required for SF reorientation in fibroblasts, mathematical modeling suggests that the reorientation of SF follows focal-adhesion slipping and a subsequent SF rotation perpendicular to the direction of stretch (Chen *et al.*, 2012; Burridge and Wittchen, 2013). ECM rigidity influences SF formation, as the level of cell-generated contractility depends on the resistance offered by the ECM at sites of adhesion. Although the level of myosin activity in response to ECM rigidity might not depend on RhoA during early adhesion (Mitrossilis *et al.*, 2010), RhoA is required for SF formation and adhesion maturation (Schiller *et al.*, 2013). RhoA affects adhesion maturation via a cooperative effect between ROCK and mDia. Initially, ROCK-stimulated myosin activity was shown to trigger adhesion maturation (Chrzanowska-Wodnicka and Burridge, 1996). When mechanical tension is applied externally to integrin-based adhesions (Riveline *et al.*, 2001), an additional RhoA effector, mDia1, is required to induce adhesion growth. This indicates that cell-generated contractility results in positive feedback to the RhoA/mDia pathway which, in turn, promotes adhesion growth through actin polymerization. Consistent with this, cell-generated tension induces the recruitment of mDia1 to adhesion complexes (Kuo *et al.*, 2011; Schiller *et al.*, 2013). mDia recruitment and activation contributes to adhesion complex reinforcement and maturation, which can be stimulated either by external tension or ROCK-mediated contractility.

At cadherin-based adhesions, Rac1 initiates actin assembly through Arp2/3 and WAVE2 (WASP-family verprolin homologous protein 2), whereas RhoA activates ROCK1 to stimulate myosin II-mediated cortical contractility (Priya *et al.*, 2017). Additionally, RhoA activates mDia and promotes the formation of stable actin bundles that maintain the integrity of cell–cell junction subjected to sustained tension (Acharya *et al.*, 2017). Interestingly, neighboring cells apply tensional forces at cell–cell junctions, which in turn promotes cadherin-based adhesion growth in a Rac1-dependent manner, potentially by stimulating actin polymerization (Liu *et al.*, 2010).

During cell migration, Rac1 drives protrusion formation by stimulating actin polymerization at the leading edge. Rac1-mediated protrusion formation generates an increase in membrane tension that propagates and maintains cell polarity by restricting Rac1 activation and preventing the initiation of additional protrusions (Houk *et al.*, 2012). During collective migration, Rho proteins and mechanical stress transmission also play a role in orchestrating multicellular organization and polarity. In the leader cell, RhoA-mediated ROCK activity produces high traction forces on the substrate that drag the multicellular structure (Reffay *et al.*, 2014). In follower cells, Rac1 and merlin are involved in a negative feedback that regulate Rac1-mediated lamellipod formation in response to tension at cell–cell junctions (Das *et al.*, 2015).

5.4.2. *Proliferation and cell survival*

Cyclic stretch stimulates proliferation in different cells types, but the mechanisms and the Rho proteins involved are different. In smooth muscle cells (SMCs), RhoA is activated in response to cyclic stretch and stimulates proliferation through a ROCK-dependent and mDia-mediated effect on p27, leading to G1/S progression (Jaffe and Hall, 2005). In endothelial cells, cyclic stretch generates tension on VE-cadherin-based adhesions that activates Rac1 (Liu *et al.*, 2007). Once activated, Rac1 stimulates proliferation by increasing cyclin D1 expression (Jaffe and Hall, 2005).

Cells cultured on rigid substrate proliferate more rapidly. Surprisingly, Rac1 is necessary for this pro-proliferative effect (Bae *et al.*, 2014). A recent study determined that ECM stiffness triggers a FAK and p130Cas-dependent mechanism that promotes Rac activity (Bae *et al.*, 2014). The enhanced Rac activity led to a sustained increase in intracellular stiffness, induction of cyclin D1, and S phase entry.

Interestingly, the Rac signaling creates a positive feedback loop where Rac-dependent cell stiffening further perpetuates FAK-p130 Cas signaling (Bae *et al.*, 2014).

Rac1 also regulates apoptosis in response to mechanical stress. Depletion of FilGAP/ARHGAP24 increases apoptosis in response to tensional forces applied to adhesions (Shifrin *et al.*, 2009). This indicates that FilGAP-mediated Rac1 inhibition serves as a mechanotransduction mechanism that promotes cell survival in cells subjected to mechanical stress.

5.4.3. *Transcription and differentiation*

RhoA can regulate the activity of transcription factors indirectly by controlling the organization of the actin cytoskeleton. In response to tensional forces applied to integrin-based adhesions, RhoA activates ROCK and mDia, which promote actin remodeling and SF formation, leading to the nuclear translocation of myocardin-related transcription factor-A (MRTF-A) and serum response factor (SRF)-dependent gene transcription (Chan *et al.*, 2010a, b). RhoA-mediated SRF activation in response to force contributes to myofibroblast differentiation (Chan *et al.*, 2010b). Changes in ECM rigidity are a potent regulator of gene expression, sufficient to direct mesenchymal stem cell differentiation. When cells are cultured on rigid substrates, myosin contractility induces the nuclear translocation of Yes-associated protein (YAP) and TAZ (transcriptional coactivator with PDZ-binding motif), two transcription co-activators. By increasing ROCK-dependent contractility, RhoA activation in response to ECM rigidity contributes to YAP nuclear translocation and subsequent induction of gene expression (Dupont *et al.*, 2011), thus mediating the effect of ECM stiffness on cell differentiation.

By stimulating the activity of NADPH oxidase, Rac1 regulates the production of reactive oxygen species (ROS) in many cell types (Jaffe and Hall, 2005). In endothelial cells subjected to shear stress, Rac1 induces ROS production, leading to NF-κB activation and the expression of proinflammatory molecules, such as ICAM-1 (Tzima *et al.*, 2002).

5.5. Rho signaling, mechanotransduction, and diseases

5.5.1. *Rho mechanoresponse and cancer*

Since RhoA and Rac1 regulate many cellular processes important for tumorigenesis, the involvement of Rho GTPases during cancer development has prompted many studies, whose results have been partially contradictory (Sahai and Marshall, 2002). During the development of solid tumors, tissue architecture is extensively modified and displays abnormal physical properties. ECM becomes stiffer, offering stronger resistance to the tension generated by the cells. In turn, the cells respond by increased contractility, developing isometric force and restoring tensional equilibrium. The resulting excessive cytoskeletal tension contributes to the altered behavior of malignant cells (Ghosh *et al.*, 2008; Paszek *et al.*, 2005) Increase in cell-generated force is associated with excessive FAK signaling, ROCK-mediated disruption of adherens junctions and ERK activation, which promote tumor cell proliferation (Paszek *et al.*, 2005). Interestingly, inhibition of ROCK or ERK signaling results in a significant decrease in tumor cell proliferation and repression of the malignant phenotype (Paszek *et al.*, 2005). RhoA activation in response to the altered microenvironment might also contribute to metastasis by activation of MRTF-A-mediated transcription regulation. Depletion of MRTF-A or SRF reduces cell invasion and motility in culture and *in vivo* (Medjkane *et al.*, 2009), indicating that RhoA-mediated MRTF-A activation promotes cancer metastasis.

5.5.2. *Rho mechanoresponse and cardiovascular diseases*

Vascular system development and remodeling are extremely sensitive to changes in blood pressure and blood flow. SMCs respond to acute changes in blood pressure through RhoA-mediated changes in myogenic tone (Hahn and Schwartz, 2009). When this increase in pressure is maintained, which is the case during hypertension, stretch-induced RhoA/ROCK activation promotes SMC proliferation (Liu *et al.*, 2007), leading to blood vessel remodeling and arterial wall thickening. Changes in blood flow impact endothelial cells. Low and oscillatory shear stress promotes inflammation and atherosclerosis development, while high and unidirectional shear stress prevents inflammation and plaque formation. Rac1-dependent NF-κB activation contributes to flow-induced increase in ICAM-1 expression and recruitment of leukocytes to atherosclerotic plaque (Tzima, 2006).

5.6. Perspectives

It becomes evident that RhoA and Rac1 have a central position in mechanotransduction, as they contribute to every step from force generation, propagation to transduction. Similar to biochemical signaling cascade, mechanical stress transmission and transduction via Rho proteins allow not only molecular process coordination at the cellular scale but also intercellular communication. The development of innovative tools and methods in biophysics and system biology will increase our understanding of the complex cross talk that mechanical tension and Rho proteins operate in cells and may reveal that Rho effector or regulator act directly as mechanosensor protein.

References

Abiko, H., Fujiwara, S., Ohashi, K., Hiatari, R., Mashiko, T., Sakamoto, N., Sato, M., and Mizuno, K. (2015). Rho guanine nucleotide exchange factors involved in cyclic-stretch-induced reorientation of vascular endothelial cells. J. Cell Sci. *128*, 1683–1695.

Acharya, B.R., Wu, S.K., Lieu, Z.Z., Parton, R.G., Grill, S.W., Bershadsky, A.D., Gomez, G.A., and Yap, A.S. (2017). Mammalian diaphanous 1 mediates a pathway for E-cadherin to stabilize epithelial barriers through junctional contractility. Cell Rep. *18*, 2854–2867.

Bae, Y.H., Mui, K.L., Hsu, B.Y., Liu, S.-L., Cretu, A., Razinia, Z., Xu, T., Puré, E., and Assoian, R.K. (2014). A FAK-Cas-Rac-lamellipodin signaling module transduces extracellular matrix stiffness into mechanosensitive cell cycling. Sci. Signal. *7*, ra57.

Balaban, N.Q., Schwarz, U.S., Riveline, D., Goichberg, P., Tzur, G., Sabanay, I., Mahalu, D., Safran, S., Bershadsky, A., Addadi, L., et al. (2001). Force and focal adhesion assembly: a close relationship studied using elastic micropatterned substrates. Nat. Cell Biol. *3*, 466–472.

Burridge, K. and Wittchen, E.S. (2013). The tension mounts: stress fibers as force-generating mechanotransducers. J. Cell Biol. *200*, 9–19.

Cai, D., Chen, S.-C., Prasad, M., He, L., Wang, X., Choesmel-Cadamuro, V., Sawyer, J.K., Danuser, G., and Montell, D.J. (2014). Mechanical feedback through E-cadherin promotes direction sensing during collective cell migration. Cell *157*, 1146–1159.

Chan, M.W.C., Chaudary, F., Lee, W., Copeland, J.W., and McCulloch, C.A. (2010a). Force-induced myofibroblast differentiation through collagen receptors is dependent on mammalian diaphanous (mDia). J. Biol. Chem. *285*, 9273–9281.

Chan, M.W.C., Chaudary, F., Lee, W., Copeland, J.W., and McCulloch, C.A. (2010b). Force-induced myofibroblast differentiation through collagen receptors is dependent on mammalian diaphanous (mDia). J. Biol. Chem. *285*, 9273–9281.

Chen, B., Kemkemer, R., Deibler, M., Spatz, J., and Gao, H. (2012). Cyclic stretch induces cell reorientation on substrates by destabilizing catch bonds in focal adhesions. PLos ONE. *7*: e48346.

Choquet, D., Felsenfeld, D.P., and Sheetz, M.P. (1997). Extracellular matrix rigidity causes strengthening of integrin-cytoskeleton linkages. Cell *88*, 39–48.

Chrzanowska-Wodnicka, M. and Burridge, K. (1996). Rho-stimulated contractility drives the formation of stress fibers and focal adhesions. J. Cell Biol. *133*, 1403–1415.

Collins, C., Guilluy, C., Welch, C., O'Brien, E.T., Hahn, K., Superfine, R., Burridge, K., and Tzima, E. (2012). Localized tensional forces on PECAM-1 elicit a global mechanotransduction response via the integrin-RhoA pathway. Curr. Biol. *22*, 2087–2094.

Das, T., Safferling, K., Rausch, S., Grabe, N., Boehm, H., and Spatz, J.P. (2015). A molecular mechanotransduction pathway regulates collective migration of epithelial cells. Nat. Cell Biol. *17*, 276–287.

Dupont, S., Morsut, L., Aragona, M., Enzo, E., Giulitti, S., Cordenonsi, M., Zanconato, F., Le Digabel, J., Forcato, M., Bicciato, S., et al. (2011). Role of YAP/TAZ in mechanotransduction. Nature *474*, 179–183.

Echarri, A. and Del Pozo, M.A. (2015). Caveolae — mechanosensitive membrane invaginations linked to actin filaments. J. Cell Sci. *128*, 2747–2758.

Garcia-Mata, R., Boulter, E., and Burridge, K. (2011). The "invisible hand": regulation of RHO GTPases by RHOGDIs. Nat. Rev. Mol. Cell Biol. *12*, 493–504.

Gawlak, G., Tian, Y., O'Donnell, J.J., Tian, X., Birukova, A.A., and Birukov, K.G. (2014). Paxillin mediates stretch-induced Rho signaling and endothelial permeability via assembly of paxillin-p42/44MAPK-GEF-H1 complex. FASEB J. *28*, 3249–3260.

Geiger, B., Spatz, J.P., and Bershadsky, A.D. (2009). Environmental sensing through focal adhesions. Nat. Rev. Mol. Cell Biol. *10*, 21–33.

Ghosh, K., Thodeti, C.K., Dudley, A.C., Mammoto, A., Klagsbrun, M., and Ingber, D.E. (2008). Tumor-derived endothelial cells exhibit aberrant Rho-mediated mechanosensing and abnormal angiogenesis in vitro. Proc. Natl. Acad. Sci. USA *105*, 11305–11310.

Grashoff, C., Hoffman, B.D., Brenner, M.D., Zhou, R., Parsons, M., Yang, M.T., McLean, M.A., Sligar, S.G., Chen, C.S., Ha, T., et al. (2010). Measuring mechanical tension across vinculin reveals regulation of focal adhesion dynamics. Nature *466*, 263–266.

Guilluy, C., Swaminathan, V., Garcia-Mata, R., O'Brien, E.T., Superfine, R., and Burridge, K. (2011). The Rho GEFs LARG and GEF-H1 regulate the mechanical response to force on integrins. Nat. Cell Biol. *13*, 722–727.

Hahn, C. and Schwartz, M.A. (2009). Mechanotransduction in vascular physiology and atherogenesis. Nat. Rev. Mol. Cell Biol. *10*, 53–62.

Harris, A.K., Wild, P., and Stopak, D. (1980). Silicone rubber substrata: a new wrinkle in the study of cell locomotion. Science *208*, 177–179.

Hayakawa, K., Sato, N., and Obinata, T. (2001). Dynamic reorientation of cultured cells and stress fibers under mechanical stress from periodic stretching. Exp. Cell Res. *268*, 104–114.

Heck, J.N., Ponik, S.M., Garcia-Mendoza, M.G., Pehlke, C.A., Inman, D.R., Eliceiri, K.W., and Keely, P.J. (2012). Microtubules regulate GEF-H1 in response to extracellular matrix stiffness. Mol. Biol. Cell *23*, 2583–2592.

Hoffman, B.D., Grashoff, C., and Schwartz, M.A. (2011). Dynamic molecular processes mediate cellular mechanotransduction. Nature *475*, 316–323.

Houk, A.R. Jilkine, A., Mejean, C.O., Boltyanskiy, R., Dufresne, E.R., Angenent, S.B., Altschuler, S.J., Wu, L.F., and Weiner, O.D. (2012). Membrane tension maintains cell polarity by confining signals to the leading edge during neutrophil migration. Cell *148*, 175–188.

Jaffe, A.B. and Hall, A. (2005). Rho GTPases: biochemistry and biology. Annu. Rev. Cell Dev. Biol. *21*, 247–269.

Kang, H., Liu, J., Sun, A., Liu, X., Fan, Y., and Deng, X. (2017). Vascular smooth muscle cell glycocalyx mediates shear stress-induced contractile responses via a Rho kinase (ROCK)-myosin light chain phosphatase (MLCP) pathway. Sci. Rep. *7*, 42092.

Katsumi, A., Milanini, J., Kiosses, W.B., del Pozo, M.A., Kaunas, R., Chien, S., Hahn, K.M., and Schwartz, M.A. (2002). Effects of cell tension on the small GTPase Rac. J. Cell Biol. *158*, 153–164.

Kawamura, S., Miyamoto, S., and Brown, J.H. (2003). Initiation and transduction of stretch-induced RhoA and Rac1 activation through caveolae: cytoskeletal regulation of ERK translocation. J. Biol. Chem. *278*, 31111–31117.

Kim, T.-J., Zheng, S., Sun, J., Muhamed, I., Wu, J., Lei, L., Kong, X., Leckband, D.E., and Wang, Y. (2015). Dynamic visualization of α-catenin reveals rapid, reversible conformation switching between tension states. Curr. Biol. *25*, 218–224.

Kuo, J.-C., Han, X., Hsiao, C.-T., Yates, J.R., and Waterman, C.M. (2011). Analysis of the myosin-II-responsive focal adhesion proteome reveals a role for β-Pix in negative regulation of focal adhesion maturation. Nat. Cell Biol. *13*, 383–393.

Ladoux, B., Anon, E., Lambert, M., Rabodzey, A., Hersen, P., Buguin, A., Silberzan, P., and Mège, R.-M. (2010). Strength dependence of cadherin-mediated adhesions. Biophys. J. *98*, 534–542.

Lawson, C.D. and Burridge, K. (2014). The on-off relationship of Rho and Rac during integrin-mediated adhesion and cell migration. Small GTPases *5*, e27958.

Lessey-Morillon, E.C., Osborne, L.D., Monaghan-Benson, E., Guilluy, C., O'Brien, E.T., Superfine, R., and Burridge, K. (2014). The RhoA guanine nucleotide exchange factor, LARG, mediates ICAM-1-dependent mechanotransduction in endothelial cells to stimulate transendothelial migration. J. Immunol. *192*, 3390–3398.

Liu, W.F., Nelson, C.M., Tan, J.L., and Chen, C.S. (2007). Cadherins, RhoA, and Rac1 are differentially required for stretch-mediated proliferation in endothelial versus smooth muscle cells. Circ. Res. *101*, e44–e52.

Liu, Y., Collins, C., Kiosses, W.B., Murray, A.M., Joshi, M., Shepherd, T.R., Fuentes, E.J., and Tzima, E. (2013). A novel pathway spatiotemporally activates Rac1 and redox signaling in response to fluid shear stress. J. Cell Biol. *201*, 863–873.

Liu, Z., Tan, J.L., Cohen, D.M., Yang, M.T., Sniadecki, N.J., Ruiz, S.A., Nelson, C.M., and Chen, C.S. (2010). Mechanical tugging force regulates the size of cell-cell junctions. Proc. Natl. Acad. Sci. USA *107*, 9944–9949.

Maekawa, M., Ishizaki, T., Boku, S., Watanabe, N., Fujita, A., Iwamatsu, A., Obinata, T., Ohashi, K., Mizuno, K., and Narumiya, S. (1999). Signaling from Rho to the actin cytoskeleton through protein kinases ROCK and LIM-kinase. Science *285*, 895–898.

Marjoram, R.J., Guilluy, C., and Burridge, K. (2015). Using magnets and magnetic beads to dissect signaling pathways activated by mechanical tension applied to cells. Methods *94*, 19–26.

Matthews, B.D., Overby, D.R., Mannix, R., and Ingber, D.E. (2006). Cellular adaptation to mechanical stress: role of integrins, Rho, cytoskeletal tension and mechanosensitive ion channels. J. Cell Sci. *119*, 508–518.

Medjkane, S., Perez-Sanchez, C., Gaggioli, C., Sahai, E., and Treisman, R. (2009). Myocardin-related transcription factors and SRF are required for cytoskeletal dynamics and experimental metastasis. Nat. Cell Biol. *11*, 257–268.

Mitrossilis, D., Fouchard, J., Pereira, D., Postic, F., Richert, A., Saint-Jean, M., and Asnacios, A. (2010). Real-time single-cell response to stiffness. Proc. Natl. Acad. Sci. USA *107*, 16518–16523.

Mogilner, A. and Oster, G. (2003). Force generation by actin polymerization II: the elastic ratchet and tethered filaments. Biophys. J. *84*, 1591–1605.

Paszek, M.J., Zahir, N., Johnson, K.R., Lakins, J.N., Rozenberg, G.I., Gefen, A., Reinhart-King, C.A., Margulies, S.S., Dembo, M., Boettiger, D., *et al.* (2005). Tensional homeostasis and the malignant phenotype. Cancer Cell *8*, 241–254.

Peng, F., Wu, D., Ingram, A.J., Zhang, B., Gao, B., and Krepinsky, J.C. (2007). RhoA activation in mesangial cells by mechanical strain depends on caveolae and caveolin-1 interaction. J. Am. Soc. Nephrol. JASN *18*, 189–198.

Peng, F., Zhang, B., Ingram, A.J., Gao, B., Zhang, Y., and Krepinsky, J.C. (2010). Mechanical stretch-induced RhoA activation is mediated by the RhoGEF Vav2 in mesangial cells. Cell. Signal. *22*, 34–40.

Priya, R., Liang, X., Teo, J.L., Duszyc, K., Yap, A.S., and Gomez, G.A. (2017). ROCK1 but not ROCK2 contributes to RhoA signaling and NMIIA-mediated contractility at the epithelial zonula adherens. Mol. Biol. Cell *28*, 12–20.

Reffay, M., Parrini, M.C., Cochet-Escartin, O., Ladoux, B., Buguin, A., Coscoy, S., Amblard, F., Camonis, J., and Silberzan, P. (2014). Interplay of RhoA and mechanical forces in collective cell migration driven by leader cells. Nat. Cell Biol. *16*, 217–223.

Ridley, A.J. and Hall, A. (1992). The small GTP-binding protein rho regulates the assembly of focal adhesions and actin stress fibers in response to growth factors. Cell *70*, 389–399.

Riveline, D., Zamir, E., Balaban, N.Q., Schwarz, U.S., Ishizaki, T., Narumiya, S., Kam, Z., Geiger, B., and Bershadsky, A.D. (2001). Focal contacts as mechanosensors: externally applied local mechanical force induces growth of focal contacts by an mDia1-dependent and ROCK-independent mechanism. J. Cell Biol. *153*, 1175–1186.

Sahai, E. and Marshall, C.J. (2002). RHO-GTPases and cancer. Nat. Rev. Cancer *2*, 133–142.

Schiller, H.B., Hermann, M.-R., Polleux, J., Vignaud, T., Zanivan, S., Friedel, C.C., Sun, Z., Raducanu, A., Gottschalk, K.-E., Théry, M., *et al.* (2013). β1- and αv-class integrins cooperate to regulate myosin II during rigidity sensing of fibronectin-based microenvironments. Nat. Cell Biol. *15*, 625–636.

Scott, D.W., Tolbert, C.E., and Burridge, K. (2016). Tension on JAM-A activates RhoA via GEF-H1 and p115 RhoGEF. Mol. Biol. Cell *27*, 1420–1430.

Shifrin, Y., Arora, P.D., Ohta, Y., Calderwood, D.A., and McCulloch, C.A. (2009). The role of FilGAP-filamin A interactions in mechanoprotection. Mol. Biol. Cell *20*, 1269–1279.

Tzima, E. (2006). Role of small GTPases in endothelial cytoskeletal dynamics and the shear stress response. Circ. Res. *98*, 176–185.

Tzima, E., Del Pozo, M.A., Kiosses, W.B., Mohamed, S.A., Li, S., Chien, S., and Schwartz, M.A. (2002). Activation of Rac1 by shear stress in endothelial cells mediates both cytoskeletal reorganization and effects on gene expression. EMBO J. *21*, 6791–6800.

Tzima, E., Irani-Tehrani, M., Kiosses, W.B., Dejana, E., Schultz, D.A., Engelhardt, B., Cao, G., DeLisser, H., and Schwartz, M.A. (2005). A mechanosensory complex that mediates the endothelial cell response to fluid shear stress. Nature *437*, 426–431.

Vicente-Manzanares, M., Ma, X., Adelstein, R.S., and Horwitz, A.R. (2009). Non-muscle myosin II takes centre stage in cell adhesion and migration. Nat. Rev. Mol. Cell Biol. *10*, 778–790.

Volkers, L., Mechioukhi, Y., and Coste, B. (2015). Piezo channels: from structure to function. Pflugers Arch. *467*, 95–99.

Wang, N., Butler, J.P., and Ingber, D.E. (1993). Mechanotransduction across the cell surface and through the cytoskeleton. Science *260*, 1124–1127.

Watanabe, N., Kato, T., Fujita, A., Ishizaki, T., and Narumiya, S. (1999). Cooperation between mDia1 and ROCK in Rho-induced actin reorganization. Nat. Cell Biol. *1*, 136–143.

Wong, A.J., Pollard, T.D., and Herman, I.M. (1983). Actin filament stress fibers in vascular endothelial cells *in vivo*. Science *219*, 867–869.

Zhao, X.-H., Laschinger, C., Arora, P., Szászi, K., Kapus, A., and McCulloch, C.A. (2007). Force activates smooth muscle alpha-actin promoter activity through the Rho signaling pathway. J. Cell Sci. *120*, 1801–1809.

Posttranslational modifications of Rho GTPases mediated by bacterial toxins and cellular systems

6

Emmanuel Lemichez[*,†,§], *Amel Mettouchi*[*,†], *and Michel Robert Popoff*[‡]

*Université Nice Côte d'Azur, Inserm, U1065, C3M, 151 route Saint Antoine de Ginestière, BP 2 3194, Nice 06204 Cedex 3, France

†Equipe labélisée ligue contre le cancer

‡Department of Microbiology, Bacterial Toxins Unit, Institut Pasteur, Paris 75015, France

§emmanuel.lemichez@inserm.fr

Keywords: Bacteria, toxins, actin cytoskeleton, vesicle trafficking, small GTPases, Rho, Rab, Ras, posttranslational modification.

6.1. Introduction

Extensive studies carried on Rho proteins have unveiled several layers of regulations of these small GTPases including posttranslational modifications (PTMs) (Hodge and Ridley, 2016; Popoff, 2014; Aktories, 2011). Rho GTPases are molecular switches that undergo a GTP/GDP-based cycle that is described in other chapters and reviewed in (Hodge and Ridley, 2016). Briefly, they bind and hydrolyze the guanosine triphosphate (GTP) into guanosine diphosphate (GDP) thereby oscillating between a GTP-bound active form and a GDP-bound inactive form (Wittinghofer and Vetter, 2011). The switch imposes conformational changes into two flexible loops or switch regions that are involved in effector binding and GTP hydrolysis. Transitions between both forms are controlled by guanine nucleotide exchange factors (GEFs), GTPase activating proteins (GAPs), and guanine nucleotide dissociation inhibitors (GDIs) protein families (see Chapter 3 by Amin and Ahmadian). The GTP-bound form can contact several effector proteins at the interface of membranes to coordinate the spatiotemporal organization of protein complexes that transduce signals. A series of PTMs of the CAXX-box terminal part are essential to the function of small GTPases (Olson, 2016). This stabilizes the anchorage of the GTP-bound active form in membranes. For Rho GTPases, this commonly involves a first step of geranylgeranylation of the cysteine residue of the CAXX motif, depicted in Fig. 1, that is followed by the endoproteolytic removal of the last three amino-acids and subsequent carboxymethylation of the cysteine residue; see Liu *et al.* (2012) for review. In addition, a reversible palmitoylation of the Cys-178, located up-stream of the CAXX-motif, in Rac1 targets this GTPase to cholesterol–rich liquid ordered membrane microdomains, where it is stabilized in its active form to control actin cytoskeleton dynamic and thereby membrane organization (Navarro-Lérida *et al.*, 2012).

Rho proteins are essential signaling hubs that belong to the superfamily of p21 Ras-related small GTPases. They share a high degree of amino-acid sequence identity that witnesses a high selective pressure on critical amino acids directly involved in the binding and hydrolysis of the guanine nucleotides, as well as their control of essential cellular functions (Olson, 2016; Hodge and Ridley, 2016). Apart from the highly conserved GTPase scaffold, Rho GTPase members display specific amino-acid sequences in particular in the switch-I region and at their carboxy-terminal end. In addition, at the difference of other small GTPases, Rho proteins display an insert region (amino acids 124–136 in RhoA). These variations of primary amino-acid sequence confer to Rho proteins their identity of interaction with regulators and effectors, as well as their specificity of localization at the membrane interface (Olson, 2016; Hodge and Ridley, 2016). In humans, there are 20 Rho GTPase members that can be subdivided into four evolutionary conserved clusters, comprising Rac, Rho, RhoH, and RhoBTB (Olson, 2016; Boureux *et al.*, 2007) (see Chapters 1 and 2 by Fort and Aspenström). The three most studied Rho proteins, RhoA, Rac1, and Cdc42, are critical regulators of the architecture and dynamics of the actin cytoskeleton that sculpts cell shape and provides contractile forces, thereby conferring to the cells their capacity to adhere to the matrix, form junctions in cell monolayers, as well as migrate and phagocyte microbes (Ridley and Hall, 1992; Ridley *et al.*, 1992; Nobes and Hall, 1995; Caron and Hall, 1998). Remarkably, Rho members are the target of PTMs catalyzed by an impressive array of bacterial and cellular factors discussed in this chapter.

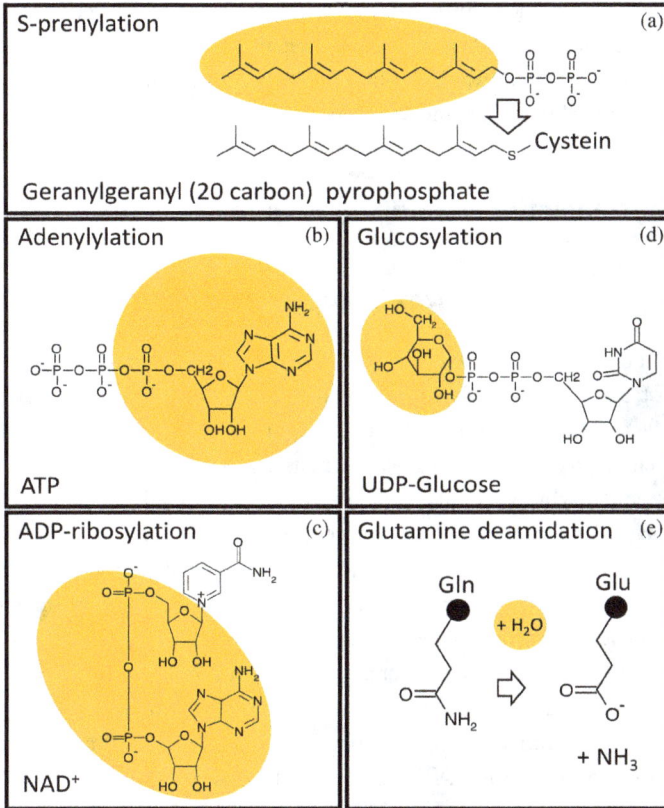

Figure 1. Examples of cosubtrates involved in PTMs. The figure depicts in orange the part of the molecule coupled to amino acid residues on the target: (a) This example of S-prenylation corresponds to the cross-linking of the 20-carbon backbone of geranylgeranyl to the cysteine residue of the CAXX-box motif. (b), (c), (d) ATP, NAD⁺, and UDP-glucose cosubtrates involved in reactions of adenylylation (or AMPylation), mono-ADP-ribosylation, and glucosylation, respectively. (e) Reaction of glutamine deamidation.

Deconstructing our understanding of the regulation and functions of small Rho GTPases by means of bacterial toxin investigations is a strategy that provides an invaluable source of information on these molecular switches (Visvikis *et al.*, 2010; Lemichez and Aktories, 2013). This prolific research in cellular microbiology has unveiled an unprecedented convergence of bacterial virulence toward host GTPases notably of the Rho family. The study of *C. botulinum* C3 exoenzyme was the pioneer investigation of Rho protein in the control of actin cytoskeleton (Chardin *et al.*, 1989; Paterson *et al.*, 1990). Here, we review the intimate cross talk between bacteria and Rho proteins. This comprises newly described PTMs of small GTPases other than Rho proteins given that it broadens our knowledge of the repertoire of possibilities. In addition, we discuss examples of PTMs of Rho GTPases by the cellular machinery.

6.2. Post-translational modifications of small Rho GTPases

6.2.1. Bacterial effector-mediated posttranslational modifications of Rho GTPases

Modulation of Rho GTPase activity by PTMs is a strategy exploited by bacteria to breach host barriers and defenses (Lemichez and Aktories, 2013). They produce virulence factors that are either AB-type toxins, endowed with the capacity to enter cells via endocytosis and translocate their enzymatic part into the cytosol (Lemichez and Barbieri, 2013), or factors directly injected through syringe-like secretion systems into the cytosol by cell-bound bacteria, here referred to as injected effectors (Galan, 2009). PTMs of Rho proteins catalyzed by bacterial factors irreversibly activate or inactivate small GTPases and promote major pathophysiological outcomes (Lemichez and Aktories, 2013). The unregulated activation or inhibition of Rho proteins by bacterial factors can produce similar effects such as efficient disruption of epithelial and endothelial barriers (Lemichez and Aktories, 2013). Another example is provided by recent findings showing that bacteria-induced inhibition of RhoA or activation of Rac1 both trigger Caspase-1-mediated processing of pro-IL-1β (Xu *et al.*, 2014; Aubert *et al.*, 2016; Eitel *et al.*, 2012; Diabate *et al.*, 2015; Zhao and Shao, 2016). On the other hand, the bidirectional modulation of Rho GTPase activity by PTMs can produce antagonist effects. For example, although the activation of Rho proteins, notably Rac1, by Gram-negative enteric bacteria foster their entry into epithelial cells by conferring to epithelial cells phagocytic-like properties (Boquet and Lemichez, 2003), the inhibition of Rho proteins is a strategy used by *Yersinia* spp. or *Pseudomonas* to block phagocytosis (Popoff, 2014; Caron and Hall, 1998). Thus, the bidirectional modulation of Rho protein activities combined with a targeting of different combinations of small GTPases confer general and specific pathogenic features to bacteria.

A large repertoire of PTMs of Rho GTPases has been characterized over the past 20 years, clarifying the mode of action of essential virulence factors from several major human pathogens, such as *Clostridium* spp., *Yersinia* spp. and *Escherichia coli* (Popoff, 2014; Lemichez and Aktories, 2013; Aktories, 2011) (Table 1). As discussed below and reviewed in Aktories (2011), these PTMs

Table 1. Examples of PTMs of small GTPases induced by bacterial factors. Q: glutamine (Gln); T: threonine (Thr); S: serine (Ser); C: cysteine (Cys); N: asparagine (Asn).

PTMs	Modified amino-acids & targets	Virulence factors	References
Deamidation	Q63 in RhoA (Q61 in Rac1/Cdc42)	CNF1 from *E. coli*	Flatau *et al.* (1997); Schmidt *et al.* (1997)
	N41 in RhoA (N39 in Rac 1)	TecA from *B. cenocepacia*	Aubert *et al.* (2016)
Transglutamination	Q63 in RhoA (Q61 in Rac1/Cdc42)	DNT from *Bordetella* spp.	Masuda *et al.* (2000)
Glucosylation with UDP-glucose	T37 in RhoA (T35 in Rac1/Cdc42)	TcdA & TcdB from *C. difficile*	Just *et al.* (1995)
Glucosamination with UDP-*N*-acetylglucosamine	T37 in RhoA (T35 in Rac1/Cdc42)	α-toxin from *C. novyi*	Selzer *et al.* (1996)
AMPylation	T37 in RhoA (T35 in Rac1/Cdc42)	VopS from *V. parahaemolyticus*	Yarbrough *et al.* (2009)
Endoproteolysis	C190 in RhoA	YopT from *Yersinia* spp.	Shao *et al.* (2003)
Phosphocholination	S79 in Rab1A (S77 Rab1B)	AnkX from *L. pneumophila*	Mukherjee *et al.* (2011)
Ubiquitin cross-linking	ADP-ribosylation of ubiquitin N42 followed by phosphoribosylated-ubiquitin formation or transfer to Rab33b	SdeA from *L. pneumophila*	Qiu *et al.* (2016); Bhogaraju *et al.* (2016)

irreversibly switch on or off Rho proteins. There is only one way to short-circuit irreversibly the GTP/GDP cycle that consists in blocking the intrinsic GTPase activity. This can be achieved by PTMs of a critical glutamine (Q) residue that positions the water molecule for the nucleophilic attack of the γ-phosphate of GTP (Wittinghofer and Vetter, 2011; Aktories, 2011). PTMs go from a discrete deamidation of this glutamine residue into a glutamic acid (Flatau et al., 1997; Schmidt et al., 1997), up to the covalent attachment of bulky chemical groups, such as lysine and polyamines (Masuda et al., 2000) or the ADP-ribose moiety from NAD^+ (Lang et al., 2010) (Fig. 1). In contrast, PTMs of several critical amino-acid residues of the switch-I domain, or of the carboxy-terminal part of Rho protein, can block signal transduction mediated by Rho proteins. Modifications of the switch-I domain encompass by order of discovery mono-ADP-ribosylation (Chardin et al., 1989; Sekine et al., 1989), glucosylation (Just et al., 1995), endoproteolysis (Shao et al., 2003), and AMPylation earlier known as adenylylation (Yarbrough et al., 2009; Worby et al., 2009) (Fig. 1). Several amino-acid residues in the switch-I domain are the targets of PTMs catalyzed by bacterial effectors. This comprises Tyr-34, Thr-37, and Asn-41 of RhoA. For example, large glycosylating toxins from Clostridia target a threonine residue located at the middle of the switch-I domain of several Rho GTPases (Thr-35/37 for Rac1/RhoA) using either uridine diphosphate (UDP)-glucose or UDP-N-acetylglucosamine, as cosubtrates (Table 1). All known toxins modify only one specific amino acid of Rho but commonly several Rho members. Only, C3-like exoenzymes from several Gram-positive bacteria (Aktories, 2011) are relatively specific of a Rho GTPase, namely RhoA. Amino acids such as Asn-41 and Tyr-34 in RhoA undergo different types of PTMs catalyzed by unrelated bacterial factors. Indeed, RhoA Asn-41 can be either mono-ADP-ribosylated by C3-like exoenzymes or deaminated into an aspartic acid by TecA from Burkholderia cenocepacia (Aubert et al., 2016). Similarly, the Tyr-34 of RhoA can be either AMPylated by IbpA from Histophilus somni or glycosylated by PaTox from Photorhabdus asymbiotica, a toxin that uses UDP-N-acetylglucosamine as cosubstrate (Worby et al., 2009; Jank et al., 2013).

Taken collectively, this convergence of virulence toward a group of host proteins certainly highlights the critical function of Rho GTPases in host defenses for a broad array of host organisms and notably protozoan natural bacterial predators. Pathogens frequently combine the attack of Rho GTPases together with that of actin molecules. For example, virulent strains of C. difficile produce both a two-component toxin (CDT) that mono-ADP-ribosylates actin and large glucosylating toxins (TcdA and TcdB) that target RhoA, Rac1, and Cdc42 (Aktories et al., 2011). The Tc toxin from Photorhabdus luminens ADP-ribosylates Rho proteins and actin thereby disrupting the architecture of the actin cytoskeleton into F-actin clusters (Lang et al., 2010).

These examples suggest that the targeting of both actin and Rho GTPases likely concurs to an efficient disruption of the actin cytoskeleton. Nevertheless, a likely hypothesis is that such complementary virulence might also sign the importance of targeting several functions controlled by Rho GTPase apart from the control of actin cytoskeleton regulation.

Some pathogens inject into host cells an array of virulence effectors that display different biochemical activities toward various subsets of Rho proteins. This is probably best illustrated by *Yersinia pestis*, the agent of the bubonic plague, and other *Yersinia* species that trigger lymphadenopathy or acute enteritis. Here, virulence effectors targeting Rho proteins act in concert to freeze phagocytic function of macrophages. Indeed, *Yersinia* injects several Yop factors targeting Rho GTPases, comprising (i) the endoprotease YopT that cleaves the carboxy-terminal part of RhoA (Shao *et al.*, 2003), (ii) YopE that displays a GAP-like activity on Rho proteins, comprising RhoA (Aili *et al.*, 2006) and (iii) YpkA (YopO) that contains a Rho GDI-like domain and a kinase domain targeting heterotrimeric Gαq to produce an indirect inhibition of RhoA (Navarro *et al.*, 2007; Prehna *et al.*, 2006). Unexpectedly, inhibition of RhoA in macrophages leads in turn to an activation of pyrin inflammasome for IL-1β secretion and the induction of pyroptotic cell death (Xu *et al.*, 2014). Remarkably, the bacterium injects another effector to tamper innate immune responses that would otherwise result from pyrin inflammasome activation (Trosky *et al.*, 2008). Indeed, in nonintoxicated cells, RhoA activates the serine/threonine protein kinase C-related kinases (PRK or PKN isoforms 1 and 2) that bind to and phosphorylate pyrin to promote its trapping by 14-3-3. This sequestration keeps pyrin inflammasome silent (Park *et al.*, 2016). Interestingly, *Yersinia* spp. opposes the induction of pyrin inflammasome by coinjecting YopM (Chung *et al.*, 2014, 2016; Schoberle *et al.*, 2016). Molecularly, YopM binds to PRK to promote a RhoA-independent phosphorylation of pyrin that restores its interaction with 14-3-3 for inactivation (Aubert *et al.*, 2016; Chung *et al.*, 2016). PTM-driven activation of Rac1 by the CNF1 toxin from *E. coli* also triggers caspase-1 mediated IL-1β secretion (Diabate *et al.*, 2015). Collectively, these findings point for inflammasome complexes as sensors of RhoA and Rac1 activities.

6.2.2. *Extended diversity of small GTPase PTMs*

It is intriguing why GTPases (small, heterotrimeric, or large GTPases, such as elongation factors EF) are frequent substrates of various toxins and injected effectors (Lemichez and Barbieri, 2013). Bacterial effectors do not induce a unique type of modification but selectively modify GTPases through diverse enzymatic reactions. Novel toxin-dependent PTMs of small GTPases of the Rab subfamily have been recently uncovered (Table 1). It is interesting to briefly discuss these PTMs, considering the diverted use of enzyme reactions found in nature. Indeed, the control of small GTPases by injected virulence effectors is not a peculiarity of Rho proteins. For example, the mono-ADP-ribosyl transferase activity of ExoS from *Pseudomonas aeruginosa* inactivates several Ras small GTPases; among them are Ras, RalA, as well as Rab5 and several other Rab members (Fraylick *et al.*, 2002). The subfamily of Rab GTPases are critical regulators of membrane trafficking and thereby targets of choice for bacteria that thrive in intracellular compartments (Wandinger-Ness and Zerial, 2014; López de Armentia *et al.*, 2016). The activity of ExoS from *Pseudomonas aeruginosa* is, for example, essential for the bacterium to avoid the reaching of lytic compartments once it gets internalized into epithelial cells (Angus *et al.*, 2010). Exciting new studies unravel that bacterial effectors from *Legionella pneumophila* modify various cellular substrates including Rab GTPases, notably Rab1 family members. This intracellular pathogen injects into host cells more than 300 effectors that confer to bacteria versatile capacities to replicate in a broad number of host species, notably the fresh water amoebas, natural hosts, or humans leading to Legionnaire's disease. Bacteria replicate in a *Legionella* containing vacuole (LCV) at the membrane of which it recruits Rab1 family members in order to hijack vesicles derived from the Golgi and the endoplasmic reticulum. The injected virulence effector AnkX catalyzes a new type of PTM. This consists in the transfer of the phosphocholine moiety of cytidine diphosphorylcholine (CDP) on the Ser-79 of Rab1A (S76 in Rab1B) that is located in the switch-II domain (Mukherjee *et al.*, 2011). In the host cell, AnkX produces an enlargement of early endosomes and impairs alkaline phosphatase secretion as does expression of Rab1 S79A. In addition, Rab1 is the target of the glutamine synthetase adenylyl transferase (GS-ATase)-like domain of DrrA (SidM) that AMPylates Tyr-77 of Rab1A (Y80 in Rab1B) to promote its activation (Müller *et al.*, 2010). This indicates that several amino-acid residues in the switch-II can be modified to promote the activation of a small GTPase. Interestingly, the phospho-cholination and AMPylation of Rab1 can be reverted by the action of Lem3 and SidD effectors, respectively (Tan and Luo, 2011; Tan *et al.*, 2011). The PTMs of Rab GTPases catalyzed by injected effectors from *Legionella* contrast with known PTMs of Rho proteins that are irreversible. This highlights the importance of keeping on and off Rab1 cycling to successfully hijack its functions. This idea is reinforced by the finding that the GDI displacement and GEF domains of DrrA can work cooperatively with the GAP activity of LepB to foster the recruitment and activation of Rab1 family members to the LCV (Ingmundson *et al.*, 2007). How precisely these factors cooperate to corrupt the spatiotemporal cycling of Rab GTPases is a challenging question to be addressed. One element, brought to our comprehensive view is that the phosphocholination of Rab tames the GTP/GDP cycle by blocking the GEF activity of the host Connecdenn (Dennd1A) protein, while allowing the binding of GEF-domain of DrrA to promote Rab1 cycling to the LCV (Mukherjee *et al.*, 2011).

The repertoire of PTMs of small GTPases has been extended recently with the description of an atypical cross-linking of phosphoribosyl-ubiquitin on Rab33b. This PTM is catalyzed by the SidE effector family from *Legionella pneumophila* that comprises SdeA (Qiu *et al.*, 2016; Bhogaraju *et al.*, 2016). SdeA contains several domains notably a phosphodiesterase (PPE) and a mono–ADP-ribosyltransferase (mART). SdeA catalyzes the ADP-ribosylation of the Arg-42 of ubiquitin using NAD^+ as cosubstrate (Qiu *et al.*, 2016; Bhogaraju *et al.*, 2016) (Fig. 1). This PTM of ubiquitin is followed by the cleavage of the phosphodiesterase bond of mono–ADP-ribosylated ubiquitin to produce phosphoribosyl-Ub (pUb) or the transfer of pUb to serine residues in Rab33b (Bhogaraju *et al.*, 2016). Canonical reaction of ubiquitination involves a first step of ubiquitin activation by an E1 enzyme followed by the transfer of ubiquitin from E1 to an E2 enzyme, prior E3-mediated direct or indirect cross-linking of ubiquitin to target proteins (Swatek and Komander, 2016). The phosphoribosyl-Ub poisons all conventional transfer reactions of ubiquitin. It works in concert with the amino-terminal deubiquitinase domain DUB of SdeA to disengage host protein modulation by ubiquitination (Bhogaraju *et al.*, 2016). The freezing of the host ubiquitination machinery likely plays a dominant role, over Rab phosphoribosyl-ubiquitin modification, in term of virulence.

6.2.3. Host reaction to bacteria-mediated PTM of Rho GTPases

The study of Rho GTPase activation by the CNF1 toxin from pathogenic *Escherichia coli* has unveiled a new mode of regulation of Rho proteins by ubiquitin-mediated proteasomal degradation (Doye *et al.*, 2002; Mettouchi and Lemichez, 2012). The CNF1 toxin is produced by pathogenic strains of *E. coli* of the phylogenic group B2. These bacteria reside in the normal gut flora of healthy human carriers while promoting extraintestinal infections such as urinary tract infections and bacteremia (Welch, 2016; Buc *et al.*, 2013). The CNF1 toxin is a paradigm of deamidase toxin and virulence effector targeting Rho GTPases found in a broad array of bacterial pathogens. CNF1 catalyzes a point mutation of Rho GTPases. This consists in a deamidation of the Gln-63 into a glutamic acid (61 in Rac1 and Cdc42) (Flatau *et al.*, 1997; Schmidt *et al.*, 1997; Lerm *et al.*, 1999). Mutation of Gln-63/61 of RhoA/Rac1 impairs their intrinsic GTPase activities thereby turning on permanently Rho proteins (Flatau *et al.*, 1997; Schmidt *et al.*, 1997; Lerm *et al.*, 1999). Cells can react to this unregulated activation of Rho GTPases by promoting their polyubiquitination for targeting to degradation by the 26S proteasome (Doye *et al.*, 2002). The HECT-domain and ankyrin-repeats containing E3 ubiquitin protein ligase 1 (HACE1) catalyzes the polyubiquitination of activated Rac1 on Lys-147 (Castillo-Lluva *et al.*, 2013; Torrino *et al.*, 2011). The depletion of HACE1 leads to an increase of total and active Rac1 cellular levels in intoxicated but also nonintoxicated cells (Castillo-Lluva *et al.*, 2013; Torrino *et al.*, 2011). HACE1 is therefore a critical element in the control of (i) Rac1-dependent NADPH oxidase-mediated production of reactive oxygen species, (ii) cyclin-D1 expression and cell cycle progression and (iii) autophagic clearance of protein aggregates (Daugaard *et al.*, 2013; Rotblat *et al.*, 2014; Zhang *et al.*, 2014).

These guard functions of HACE1 on cell homeostasis and the control of Rac1-mediated cell cycle progression likely account in part for its critical tumor suppressor function (Zhang *et al.*, 2007). Moreover, HACE1 regulates the tumor necrosis factor (TNF) receptor-1 (TNFR-1) signaling by shifting the balance of TNF signaling toward apoptosis instead of necroptosis and the ensuing inflammation (Ellerbroek *et al.*, 2003). Mice knocked out for HACE1 display colitis when the intestinal epithelium barrier is compromised by treatment with the chemical irritant (DSS) and show a significant lower resistance to infection by *Listeria monocytogenes* (Tortola *et al.*, 2016). A growing number of studies have established the importance of ubiquitin and proteasome systems (UPSs) in the control of Rho protein function (Table 2). Specific degradation of Rho proteins secures the spatial control of Rho activities and adapts the cellular levels of Rho to directly impact the extent of Rho proteins entering the GEF/GAP cycle (Wang *et al.*, 2003; Chen *et al.*, 2009; Torrino *et al.*, 2011).

6.3. Cellular PTMs of Rho GTPases

Rho proteins are highly conserved from protozoan to humans. Most likely, the cellular PTMs of small GTPases represent a necessary adaptation to increasing levels of complexity found in multicellular organisms (Hodge and Ridley, 2016). Initially, Rho GTPases have been identified as membrane proximal regulators of the actin cytoskeleton in response to engagement of growth factors and G protein-coupled proteins' receptors, as well as integrin matrix-adhesion receptors (Ridley and Hall, 1992; Ridley *et al.*, 1992; Nobes and Hall, 1995). The articulation of signals between receptor-coupled kinases and small GTPases involves the phosphorylation of GEF factors and the interplay of pleckstrin homology domains with phosphoinositides (Hodge and Ridley, 2016). As exemplified in Table 2 and reviewed in Hodge and Ridley (2016), several kinases can directly phosphorylate Rho GTPases. Despite progress made, much remains to be uncovered on PTMs

Table 2. Examples of PTMs of Rho GTPases catalyzed by cellular factors.

GTPases	PTMs & amino acid targets	PTM Effects	Enzymes	References
Rac1	Phosphorylation T108	Induces nuclear translocation	ERK	Tong et al. (2013)
	Phosphorylation Y64	Inhibition	Src and FAK	Chang et al. (2011)
	Phosphorylation S71	Inhibits GTP binding	AKT	Kwon et al. (2000)
	Ubiquitylation K147	Proteasomal degradation	HACE1	Torrino et al. (2011); Castillo-Lluva et al. (2013)
	Ubiquitylation K166	Proteasomal degradation of S71 phosphorylated Rac1 (also Rac3)	SCF-FBXL19	Zhao et al. (2013)
	Sumoylation (carboxy-terminal region)	Stabilizes the GTP-bound form	PIAS3	Castillo-Lluva et al. (2010)
RhoA	Phosphorylation S188	GDI interaction & impaired proteasomal degradation	PKA, PKG, PKC (also T127), SLK	Lang et al. (1996); Ellerbroek et al. (2003); Su et al. (2013); Guilluy et al. (2008)
	Ubiquitylation K6, K7 and K51	Proteasomal degradation	Smurf1 and Smurf2	Ozdamar et al. (2005); Lu et al. (2011); Bryan et al. (2005)

(Continued)

Table 2. (*Continued*)

GTPases	PTMs & amino acid targets	PTM Effects	Enzymes	References
	Ubiquitylation	Proteasomal degradation	Cul3–BACURD	Chen et al. (2009)
	Ubiquitylation K135	Degradation	SCF–FBXL19	Wei et al. (2013)
	Serotonylation Q63 for RhoA	Activation	Transglutaminase	Walther et al. (2003)
	Nitration Y34	Activation by decreasing GDP binding	Addition of a nitro group from peroxynitrite to tyrosine residue	Rafikov et al. (2014)
Cdc42	Phosphorylation Y64	Inhibition by promoting association with GDI	Src	Tu et al. (2003)
	Phosphorylation S185	Inhibition by promoting association with GDI	PKA	Forget et al. (2002)
RhoU/ WRCH1	Phosphorylation Y254	Inhibition by relocalization to endosomal membrane	Src	Alan et al. (2010)
	Ubiquitylation K177, K248	Proteasomal degradation	Rab40A–Cullin5	Dart et al. (2015)

of Rho GTPases if one considers the catalogue of PTMs defined by proteomic that encompass phosphorylation, ubiquitination, methylation, acetylation, and succinylation (http://www.phosphosite.org).

6.3.1. Cellular and bacterial effector shared PTMs of Rho GTPases

Human Fic-domain containing huntingtin yeast-interacting protein E (HYPE) displays *in vitro* and *in vivo* adenylylation activity toward RhoA, Rac1, and Cdc42 (Worby *et al.*, 2009). The cellular activity of HYPE is probably highly regulated in cells considering that its ectopic expression does not produce a disruption of actin cytoskeleton and a rounding of cells as does the immunoglobulin-binding protein A (IbpA) from *Histophilus somni* (Worby *et al.*, 2009). Note that IbpA also AMPylates LyGDI and blocks its phosphorylation by Src tyrosine protein kinase (Yu *et al.*, 2014). Mammalian transglutaminases catalyze the serotonylation of the Gln-63 of RhoA leading to its permanent activation (Walther *et al.*, 2003). Serotonin is a neurotransmitter in the central and peripheral nervous system and a ubiquitous hormone synthetized by tryptophan hydroxylase isoenzymes to control vasoconstriction and platelet function. Mice deficient in peripheral serotonin synthesis exhibit impaired hemostasis with high risk of thrombosis and thromboembolism, while the structure of platelets is not affected. Serotonylation of RhoA and other small GTPases promotes the exocytosis of α-granules from platelets by a receptor-independent mechanism thereby favoring their adherence to tissue lesions.

6.3.2. Unique cellular PTMs of Rho GTPases

In addition to the control of GTP/GDP cycle and membrane/cytosol shuttling, cells can modulate Rho GTPases by phosphorylation (Table 2). As discussed below, phosphorylation occurs both in the carboxy-terminal part and in the switch-II domain of small GTPases thereby controlling their cellular localization and guanine nucleotide loading, respectively. Although phosphorylation can be rapidly reverted, little is known on phosphatases that contribute to this balance. In contrast, progress has been made in identifying kinases targeting Rho proteins.

Two amino-acid residues in the switch-II domain of Rac1 can be phosphorylated thereby stabilizing the GDP-bound form. These PTMs comprise the phosphorylation of Ser-71 of Rac1 mediated by AKT; and Tyr-64, by Src or FAK (Chang *et al.*, 2011; Tu *et al.*, 2003; Kwon *et al.*, 2000). The phosphorylation of Rac1 by Src modulates the spreading of cells on fibronectin (Chang *et al.*, 2011).

Expression of the phosphonull mutant Rac1-Y64F increases the pool of GTP-bound Rac1 and the spreading area of cells. Phosphorylation of amino-acid residues in the carboxy-terminal region of Rho proteins can modulate their localization. For example, the ERK-mediated phosphorylation of Rac1 on Thr-108 promotes its translocation into the nucleus, thereby compartmentalizing Rac1 away from components of its signaling cascade to down-modulate its activity (Tong *et al.*, 2013). PTMs of Rho GTPases also contribute to modify their surface of interaction with regulatory proteins and effectors. Phosphorylation of Ser-188 of RhoA, among Rho phosphorylations, provides us with such an example (Lang *et al.*, 1996) (Table 2). In several cell types, the broad signaling cyclic nucleotides cAMP or cGMP promote a PKA or PKG-dependent phosphorylation of RhoA on Ser-188 (Lang *et al.*, 1996; Ellerbroek *et al.*, 2003; Rolli-Derkinderen *et al.*, 2005). PKA also phosphorylates the equivalent Ser-185 in Cdc42 that is not present in Rac1 (Forget *et al.*, 2002). Phospho-Ser-188 can promote variations in the repertoire of activated effectors by excluding ROCK as compared to mDIA-1 and PKN (Nusser *et al.*, 2006). In addition, phosphorylation of RhoA increases its capacity to interact with GDI, thereby altering its accessibility for activation and protects RhoA from proteasomal degradation mediated by Smad ubiquitylation regulatory factor 1 (Smurf1) E3 ubiquitin ligase (Forget *et al.*, 2002; Rolli-Derkinderen *et al.*, 2005). In contrast, phosphorylation of Rac1 on Ser-71 promotes its degradation by SCF-FBXL19 (Zhao *et al.*, 2013). Together, this provides examples of cross talk between phosphorylation and ubiquitin-mediated proteasomal degradation in the regulation of Rho proteins.

While ubiquitination of Rho GTPases and notably Rac1 promotes their degradation, the SUMOylation of Rac1 carboxy-terminal-located lysines provides us with an example of cellular PTM that reinforces the stability of the GTP-bound form (Castillo-Lluva *et al.*, 2010).

It is of great importance to further define how cellular PTMs of Rho GTPases contribute to their essential function in physiology. For example, the knockdown of HACE1 predisposes mice to colonic inflammation and cancer onset during aging (Tortola *et al.*, 2016; Zhang *et al.*, 2007). Yet to be demonstrated, growing evidence support a critical role of this regulation to tone down the oncogenic and inflammatory potential of Rac1 (Daugaard *et al.*, 2013; Mettouchi and Lemichez, 2012; Goka and Lippman, 2015; Andrio *et al.*, 2017).

6.3.3. Regulation of atypical Rho GTPases by PTMs

Some members of the Rho protein family, referred to as atypical small GTPases, display high intrinsic GDP exchange activities or substitutions in the GTPase domain, which alter their capacity to hydrolyze the GTP. For these GTPases that escape to the canonical switch regulation by GEF/GAP factors, PTMs are a prominent way to regulate their signaling via subcellular compartmentalization and stability. RhoU (also known as WRCH1) provides us with an example of atypical Rho GTPase regulated both by phosphorylation and ubiquitylation. Although RhoU belongs to the Cdc42 branch of small Rho GTPases, it is thought to be constitutively active (Saras *et al.*, 2004). This critical regulator of focal-adhesion dynamics is controlled by the kinase activity of Src, which catalyzes the phosphorylation Tyr-254 (Alan *et al.*, 2010). Such modification of the carboxy-terminal part of RhoU promotes a shift in its subcellular localization from the plasma membrane to endosomal membranes, that is concomitant with a reduction of active protein levels likely through the segregation of RhoU from its GEF factors. In addition, RhoU is polyubiquitylated on Lys-177 and Lys-248 (Dart *et al.*, 2015). This involves a recruitment of Rab40A, and associated ubiquitin ligase complex, by RhoU. Overexpression of Rab40A promotes RhoU ubiquitylation and reduces its cellular level. By analogy with known homologues in *Xenopus*, control of RhoU stability by Rab40A likely involves an ElonginB/C and Cullin5 E3 ubiquitin ligase complex.

6.4. Conclusions

Despite their flexible regulation by GTP/GDP-based switches, the small Rho GTPases are controlled by PTMs that engage or disengage GTPase signaling in a dominant manner. A large spectrum of biochemical modifications of these GTPases has been unveiled going from deamidation of glutamine and arginine residues up to the addition of bulky chemical groups. These posttranslational modifications affect critical amino-acid residues of both switch-I and II domains and the carboxy-terminal region. The posttranslational modifications of Rho GTPases by virulence factors is a strategy widely exploited by bacterial pathogens to disrupt or usurp cell signaling, thereby conferring pathogens a better fitness. Cells can also react to PTMs of Rho GTPases by inducing inflammasome signaling or limit activation of these GTPases via their degradation. PTMs catalyzed by virulence effectors from *L. pneumophila* let us glimpse other types of modifications of Rho GTPases to be uncovered. This rich cross talk between bacterial virulence effectors and small GTPases continues to unveil critical elements of cell signaling especially those implicated in human diseases, such as infection, inflammation, and cancer.

Acknowledgments

We are grateful to Grégory Michel for assistance in the preparation of Fig. 1. We thank Monica Rolando for critical reading of the manuscript. We acknowledge fundings from INSERM, CNRS, the Ligue Nationale Contre le Cancer (LNCC, équipe labellisée), the Association pour la Recherche sur le Cancer (ARC-SFI20111203659 and ARC-SFI20111203671), and the "Investments for the Future" LABEX SIGNALIFE ANR-11-LABX-0028-01.

References

Aili, M., Isaksson, E.L., Hallberg, B., Wolf-Watz, H., and Rosqvist, R. (2006). Functional analysis of the YopE GTPase-activating protein (GAP) activity of Yersinia pseudotuberculosis. Cell Microbiol. *8*, 1020–1033.

Aktories, K. (2011). Bacterial protein toxins that modify host regulatory GTPases. Nat. Rev. Microbiol. *9*, 487–498.

Aktories, K., Lang, A.E., Schwan, C., and Mannherz, H.G. (2011). Actin as target for modification by bacterial protein toxins. FEBS J. *278*, 4526–4543.

Alan, J.K., Berzat, A.C., Dewar, B. J., Graves, L.M., and Cox, A.D. (2010). Regulation of the Rho family small GTPase Wrch-1/RhoU by C-terminal tyrosine phosphorylation requires Src. Mol. Cell Biol. *30*, 4324–4338.

Andrio, E., Lotte, R., Hamaoui, D., Cherfils, J., Doye, A., Daugaard, M., Sorensen, P.H., Bost, F., Ruimy, R., Mettouchi, A., and Lemichez, E. (2017). Identification of cancer-associated missense mutations in hace1 that impair cell growth control and Rac1 ubiquitylation. Sci. Rep. *7*, 44779.

Angus, A.A., Evans, D.J., Barbieri, J.T., and Fleiszig, S.M. (2010). The ADP-ribosylation domain of *Pseudomonas aeruginosa* ExoS is required for membrane bleb niche formation and bacterial survival within epithelial cells. Infect. Immun. *78*, 4500–4510.

Aubert, D.F., Xu, H., Yang, J., Shi, X., Gao, W., Li, L., Bisaro, F., Chen, S., Valvano, M.A., and Shao, F. (2016). A *Burkholderia* type VI effector deamidates Rho GTPases to activate the pyrin inflammasome and trigger inflammation. Cell Host Microbe *19*, 1–11.

Bhogaraju, S., Kalayil, S., Liu, Y., Bonn, F., Colby, T., Matic, I., and Dikic, I. (2016). Phosphoribosylation of ubiquitin promotes serine ubiquitination and impairs conventional ubiquitination. Cell *167*, 1636–1649.

Boquet, P. and Lemichez, E. (2003). Bacterial virulence factors targeting Rho GTPases: parasitism or symbiosis? Trends Cell Biol. *13*, 238–246.

Boureux, A., Vignal, E., Faure, S., and Fort, P. (2007). Evolution of the Rho family of ras-like GTPases in eukaryotes. Mol Biol Evol *24*, 203–216.

Bryan, B., Cai, Y., Wrighton, K., Wu, G., Feng, X.H., and Liu, M. (2005). Ubiquitination of RhoA by Smurf1 promotes neurite outgrowth. FEBS Lett. *579*, 1015–1019.

Buc, E., Dubois, D., Sauvanet, P., Raisch, J., Delmas, J., Darfeuille-Michaud, A., Pezet, D., and Bonnet, R. (2013). High prevalence of mucosa-associated E. coli producing cyclomodulin and genotoxin in colon cancer. PLoS One *8*, e56964.

Caron, E. and Hall, A. (1998). Identification of two distinct mechanisms of phagocytosis controlled by different Rho GTPases. Science *282*, 1717–1721.

Castillo-Lluva, S., Tan, C.T., Daugaard, M., Sorensen, P.H., and Malliri, A. (2013). The tumour suppressor HACE1 controls cell migration by regulating Rac1 degradation. Oncogene *32*, 1735–1742.

Castillo-Lluva, S., Tatham, M.H., Jones, R.C., Jaffray, E.G., Edmondson, R.D., Hay, R.T., and Malliri, A. (2010). SUMOylation of the GTPase Rac1 is required for optimal cell migration. Nat. Cell Biol. *12*, 1078–1085.

Chang, F., Lemmon, C., Lietha, D., Eck, M., and Romer, L. (2011). Tyrosine phosphorylation of Rac1: a role in regulation of cell spreading. PLoS One *6*, e28587.

Chardin, P., Boquet, P., Madaule, P., Popoff, M.R., Rubin, E.J., and Gill, D.M. (1989). The mammalian G protein rhoC is ADP-ribosylated by *Clostridium botulinum* exoenzyme C3 and affects actin microfilaments in Vero cells. EMBO J. *8*, 1087–1092.

Chen, Y., Yang, Z., Meng, M., Zhao, Y., Dong, N., Yan, H., Liu, L., Ding, M., Peng, H.B., and Shao, F. (2009). Cullin mediates degradation of RhoA through evolutionarily conserved BTB adaptors to control actin cytoskeleton structure and cell movement. Mol. Cell *35*, 841–855.

Chung, L.K., Park, Y.H., Zheng, Y., Brodsky, I.E., Hearing, P., Kastner, D.L., Chae, J.J., and Bliska, J. B. (2016). The *Yersinia* virulence factor YopM hijacks host kinases to inhibit type III effector-triggered activation of the pyrin inflammasome. Cell Host Microbe *20*, 296–306.

Chung, L.K., Philip, N.H., Schmidt, V.A., Koller, A., Strowig, T., Flavell, R.A., Brodsky, I. E., and Bliska, J. B. (2014). IQGAP1 is important for activation of caspase-1 in macrophages and is targeted by *Yersinia pestis* type III effector YopM. MBio *5*, e01402–14.

Dart, A. E., Box, G. M., Court, W., Gale, M. E., Brown, J. P., Pinder, S. E., Eccles, S. A., and Wells, C. M. (2015). PAK4 promotes kinase-independent stabilization of RhoU to modulate cell adhesion. J. Cell Biol. *211*, 863–879.

Daugaard, M., Nitsch, R., Razaghi, B., McDonald, L., Jarrar, A., Torrino, S., Castillo-Lluva, S., Rotblat, B., Li, L., Malliri, A., Lemichez, E., Mettouchi, A., Berman, J. N., Penninger, J. M., and Sorensen, P. H. (2013). Hace1 controls ROS generation of vertebrate Rac1-dependent NADPH oxidase complexes. Nat. Commun. *4*, 2180.

Diabate, M., Munro, P., Garcia, E., Jacquel, A., Michel, G., Obba, S., Goncalves, D., Luci, C., Marchetti, S., Demon, D., *et al.* (2015). Escherichia coli alpha-Hemolysin counteracts the anti-virulence innate immune response triggered by the Rho GTPase activating Toxin CNF1 during Bacteremia. PLoS Pathog. *11*, e1004732.

Doye, A., Mettouchi, A., Bossis, G., Clement, R., Buisson-Touati, C., Flatau, G., Gagnoux, L., Piechaczyk, M., Boquet, P., and Lemichez, E. (2002). CNF1 exploits the ubiquitin-proteasome machinery to restrict Rho GTPase activation for bacterial host cell invasion. Cell *111*, 553–564.

Eitel, J., Meixenberger, K., van Laak, C., Orlovski, C., Hocke, A., Schmeck, B., Hippenstiel, S., N'Guessan, P.D., Suttorp, N., and Opitz, B. (2012). Rac1 regulates the NLRP3 inflammasome which mediates IL-1beta production in *Chlamydophila pneumoniae* infected human mononuclear cells. PLoS One *7*, e30379.

Ellerbroek, S.M., Wennerberg, K., and Burridge, K. (2003). Serine phosphorylation negatively regulates RhoA in vivo. J. Biol. Chem. *278*, 19023–19031.

Flatau, G., Lemichez, E., Gauthier, M., Chardin, P., Paris, S., Fiorentini, C., and Boquet, P. (1997). Toxin-induced activation of the G protein p21 Rho by deamidation of glutamine. Nature *387*, 729–733.

Forget, M.A., Desrosiers, R.R., Gingras, D., and Beliveau, R. (2002). Phosphorylation states of Cdc42 and RhoA regulate their interactions with Rho GDP dissociation inhibitor and their extraction from biological membranes. Biochem. J. *361*, 243–254.

Fraylick, J.E., Rucks, E.A., Greene, D.M., Vincent, T.S., and Olson, J.C. (2002). Eukaryotic cell determination of ExoS ADP-ribosyltransferase substrate specificity. Biochem. Biophys. Res. Commun. *291*, 91–100.

Galan, J. E. (2009). Common themes in the design and function of bacterial effectors. Cell Host Microbe *5*, 571–579.

Goka, E.T. and Lippman, M.E. (2015). Loss of the E3 ubiquitin ligase HACE1 results in enhanced Rac1 signaling contributing to breast cancer progression. Oncogene *34*, 5395–5405.

Guilluy, C., Rolli-Derkinderen, M., Loufrani, L., Bourge, A., Henrion, D., Sabourin, L., Loirand, G., and Pacaud, P. (2008). Ste20-related kinase SLK phosphorylates Ser188 of RhoA to induce vasodilation in response to angiotensin II type 2 receptor activation. Circ. Res. *102*, 1265–1274.

Hodge, R.G. and Ridley, A.J. (2016). Regulating Rho GTPases and their regulators. Nat. Rev. Mol. Cell Biol. *17*, 496–510.

Ingmundson, A., Delprato, A., Lambright, D.G., and Roy, C.R. (2007). *Legionella pneumophila* proteins that regulate Rab1 membrane cycling. Nature *450*, 365–369.

Jank, T., Bogdanovic, X., Wirth, C., Haaf, E., Spoerner, M., Bohmer, K. E., Steinemann, M., Orth, J. H., Kalbitzer, H.R., Warscheid, B., Hunte, C., and Aktories, K. (2013). A bacterial toxin catalyzing tyrosine glycosylation of Rho and deamidation of Gq and Gi proteins. Nat. Struct. Mol. Biol. *20*, 1273–1280.

Just, I., Selzer, J., Wilm, M., von Eichel-Streiber, C., Mann, M., and Aktories, K. (1995). Glucosylation of Rho proteins by *Clostridium difficile* toxin B. Nature *375*, 500–503.

Kwon, T., Kwon, D.Y., Chun, J., Kim, J.H., and Kang, S.S. (2000). Akt protein kinase inhibits Rac1-GTP binding through phosphorylation at serine 71 of Rac1. J. Biol. Chem. *275*, 423–428.

Lang, A.E., Schmidt, G., Schlosser, A., Hey, T.D., Larrinua, I.M., Sheets, J.J., Mannherz, H.G., and Aktories, K. (2010). *Photorhabdus luminescens* toxins ADP-ribosylate actin and RhoA to force actin clustering. Science *327*, 1139–1142.

Lang, P., Gesbert, F., Delespine-Carmagnat, M., Stancou, R., Pouchelet, M., and Bertoglio, J. (1996). Protein kinase A phosphorylation of RhoA mediates the morphological and functional effects of cyclic AMP in cytotoxic lymphocytes. EMBO J. *15*, 510–519.

Lemichez, E. and Aktories, K. (2013). Hijacking of Rho GTPases during bacterial infection. Exp. Cell Res. *319*, 2329–2336.

Lemichez, E., and Barbieri, J.T. (2013). General aspects and recent advances on bacterial protein toxins. Cold Spring Harb. Perspect. Med *3*, 1–13.

Lerm, M., Selzer, J., Hoffmeyer, A., Rapp, U.R., Aktories, K., and Schmidt, G. (1999). Deamidation of Cdc42 and Rac by *Escherichia coli* cytotoxic necrotizing factor 1: activation of c-Jun N-terminal kinase in HeLa cells. Infect. Immun. *67*, 496–503.

Liu, M., Bi, F., Zhou, X., and Zheng, Y. (2012). Rho GTPase regulation by miRNAs and covalent modifications. Trends Cell Biol. *22*, 365–373.

López de Armentia, M.M., Amaya, C., and Colombo, M.I. (2016). Rab GTPases and the autophagy pathway: bacterial targets for a suitable biogenesis and trafficking of their own vacuoles. Cells *5*, E11.

Lu, K., Li, P., Zhang, M., Xing, G., Li, X., Zhou, W., Bartlam, M., Zhang, L., Rao, Z., and He, F. (2011). Pivotal role of the C2 domain of the Smurf1 ubiquitin ligase in substrate selection. J. Biol. Chem. *286*, 16861–16870.

Masuda, M., Betancourt, L., Matsuzawa, T., Kashimoto, T., Takao, T., Shimonishi, Y., and Horiguchi, Y. (2000). Activation of rho through a cross-link with polyamines catalyzed by *Bordetella* dermonecrotizing toxin. EMBO J. *19*, 521–530.

Mettouchi, A. and Lemichez, E. (2012). Ubiquitylation of active Rac1 by the E3 ubiquitin-ligase HACE1. Small GTPases *3*, 102–106.

Mukherjee, S., Liu, X., Arasaki, K., McDonough, J., Galán, J.E., and Roy, C.R. (2011). Modulation of Rab GTPase function by a protein phosphocholine transferase. Nature *477*, 103–106.

Müller, M.P., Peters, H., Blümer, J., Blankenfeldt, W., Goody, R.S., and Itzen, A. (2010). The Legionella effector protein DrrA AMPylates the membrane traffic regulator Rab1b. Science *329*, 946–949.

Navarro-Lérida, I., Sánchez-Perales, S., Calvo, M., Rentero, C., Zheng, Y., Enrich, C., and Del Pozo, M. A. (2012). A palmitoylation switch mechanism regulates Rac1 function and membrane organization. EMBO J. *31*, 534–551.

Navarro, L., Koller, A., Nordfelth, R., Wolf-Watz, H., Taylor, S., and Dixon, J.E. (2007). Identification of a molecular target for the *Yersinia* protein kinase A. Mol. Cell *26*, 465–477.

Nobes, C.D. and Hall, A. (1995). Rho, rac, and cdc42 GTPases regulate the assembly of multimolecular focal complexes associated with actin stress fibers, lamellipodia, and filopodia. Cell *81*, 53–62.

Nusser, N., Gosmanova, E., Makarova, N., Fujiwara, Y., Yang, L., Guo, F., Luo, Y., Zheng, Y., and Tigyi, G. (2006). Serine phosphorylation differentially affects RhoA binding to effectors: implications to NGF-induced neurite outgrowth. Cell Signal. *18*, 704–714.

Olson, M.F. (2016). Rho GTPases, their post-translational modifications, disease-associated mutations and pharmacological inhibitors. Small GTPases 1–13. URL: http://dx.doi.org/10.1080/21541248.2016.1218407

Ozdamar, B., Bose, R., Barrios-Rodiles, M., Wang, H.R., Zhang, Y., and Wrana, J.L. (2005). Regulation of the polarity protein Par6 by TGFbeta receptors controls epithelial cell plasticity. Science *307*, 1603–1609.

Park, Y.H., Wood, G., Kastner, D.L., and Chae, J.J. (2016). Pyrin inflammasome activation and RhoA signaling in the autoinflammatory diseases FMF and HIDS. Nat. Immunol. *17*, 914–921.

Paterson, H.F., Self, A.J., Garrett, M.D., Just, I., Aktories, K., and Hall, A. (1990). Microinjection of recombinant p21rho induces rapid changes in cell morphology. J. Cell Biol. *111*, 1001–1007.

Popoff, M.R. (2014). Bacterial factors exploit eukaryotic Rho GTPase signaling cascades to promote invasion and proliferation within their host. Small GTPases *5*, e28209.

Prehna, G., Ivanov, M.I., Bliska, J.B., and Stebbins, C.E. (2006). Yersinia virulence depends on mimicry of host Rho-family nucleotide dissociation inhibitors. Cell *126*, 869–880.

Qiu, J., Sheedlo, M.J., Yu, K., Tan, Y., Nakayasu, E.S., Das, C., Liu, X., and Luo, Z.Q. (2016). Ubiquitination independent of E1 and E2 enzymes by bacterial effectors. Nature *533*, 120–124.

Rafikov, R., Dimitropoulou, C., Aggarwal, S., Kangath, A., Gross, C., Pardo, D., Sharma, S., Jezierska-Drutel, A., Patel, V., Snead, C., *et al.*, (2014). Lipopolysaccharide-induced lung injury involves the nitration-mediated activation of RhoA. J. Biol. Chem. *289*, 4710–4722.

Ridley, A.J. and Hall, A. (1992). The small GTP-binding protein rho regulates the assembly of focal adhesions and actin stress fibers in response to growth factors. Cell *70*, 389–399.

Ridley, A.J., Paterson, H.F., Johnston, C.L., Diekmann, D., and Hall, A. (1992). The small GTP-binding protein rac regulates growth factor-induced membrane ruffling. Cell *70*, 401–410.

Rolli-Derkinderen, M., Sauzeau, V., Boyer, L., Lemichez, E., Baron, C., Henrion, D., Loirand, G., and Pacaud, P. (2005). Phosphorylation of serine 188 protects RhoA from ubiquitin/proteasome-mediated degradation in vascular smooth muscle cells. Circ. Res. *96*, 1152–1160.

Rotblat, B., Southwell, A.L., Ehrnhoefer, D.E., Skotte, N.H., Metzler, M., Franciosi, S., Leprivier, G., Somasekharan, S.P., Barokas, A., Deng, Y., *et al.* (2014). HACE1 reduces oxidative stress and mutant Huntingtin toxicity by promoting the NRF2 response. Proc. Natl. Acad. Sci. USA *111*, 3032–3037.

Saras, J., Wollberg, P., and Aspenström, P. (2004). Wrch1 is a GTPase-deficient Cdc42-like protein with unusual binding characteristics and cellular effects. Exp. Cell Res. *299*, 356–369.

Schmidt, G., Sehr, P., Wilm, M., Selzer, J., Mann, M., and Aktories, K. (1997). Gln 63 of Rho is deamidated by *Escherichia coli* cytotoxic necrotizing factor-1. Nature *387*, 725–729.

Schoberle, T.J., Chung, L.K., McPhee, J.B., Bogin, B., and Bliska, J. B. (2016). Uncovering an important role for YopJ in the inhibition of caspase-1 in activated macrophages and promoting *Yersinia pseudotuberculosis* virulence. Infect. Immun. *84*, 1062–1072.

Sekine, A., Fujiwara, M., and Narumiya, S. (1989). Asparagine residue in the rho gene product is the modification site for botulinum ADP-ribosyltransferase. J. Biol. Chem. *264*, 8602–8605.

Selzer, J., Hofmann, F., Rex, G., Wilm, M., Mann, M., Just, I., and Aktories, K. (1996). *Clostridium novyi* alpha-toxin-catalyzed incorporation of GlcNAc into Rho subfamily proteins. J. Biol. Chem. *271*, 25173–25177.

Shao, F., Vacratsis, P. O., Bao, Z., Bowers, K. E., Fierke, C. A., and Dixon, J. E. (2003). Biochemical.characterization of the *Yersinia* YopT protease: cleavage site and recognition elements in Rho GTPases. Proc. Natl. Acad. Sci. USA *100*, 904–909.

Su, T., Straight, S., Bao, L., Xie, X., Lehner, C. L., Cavey, G.S., Teknos, T.N., and Pan, Q. (2013). PKC ε phosphorylates and mediates the cell membrane localization of RhoA. ISRN Oncol. *2013*, 329063.

Swatek, K.N. and Komander, D. (2016). Ubiquitin modifications. Cell Res. *26*, 399–422.

Tan, Y., Arnold, R.J., and Luo, Z.Q. (2011). *Legionella pneumophila* regulates the small GTPase Rab1 activity by reversible phosphorylcholination. Proc. Natl. Acad. Sci. USA *108*, 21212–21217.

Tan, Y. and Luo, Z.Q. (2011). *Legionella pneumophila* SidD is a deAMPylase that modifies Rab1. Nature *475*, 506–509.

Tong, J., Li, L., Ballermann, B., and Wang, Z. (2013). Phosphorylation of Rac1 T108 by extracellular signal-regulated kinase in response to epidermal growth factor: a novel mechanism to regulate Rac1 function. Mol. Cell Biol. *33*, 4538–4551.

Torrino, S., Visvikis, O., Doye, A., Boyer, L., Stefani, C., Munro, P., Bertoglio, J., Gacon, G., Mettouchi, A., and Lemichez, E. (2011). The E3 ubiquitin-ligase HACE1 catalyzes the ubiquitylation of active Rac1. Dev. Cell *21*, 959–965.

Tortola, L., Nitsch, R., Bertrand, M.J., Kogler, M., Redouane, Y., Kozieradzki, I., Uribesalgo, I., Fennell, L.M., Daugaard, M., Klug, H., *et al*. (2016). The tumor suppressor Hace1 is a critical regulator of TNFR1-mediated cell fate. Cell Rep. *15*, 1481–1492.

Trosky, J.E., Liverman, A.D., and Orth, K. (2008). *Yersinia* outer proteins: Yops. Cell Microbiol. *10*, 557–565.

Tu, S., Wu, W.J., Wang, J., and Cerione, R.A. (2003). Epidermal growth factor-dependent regulation of Cdc42 is mediated by the Src tyrosine kinase. J. Biol. Chem. *278*, 49293–49300.

Visvikis, O., Maddugoda, M. P., and Lemichez, E. (2010). Direct modifications of Rho proteins: deconstructing GTPase regulation. Biol. Cell *102*, 377–389.

Walther, D.J., Peter, J.U., Winter, S., Holtje, M., Paulmann, N., Grohmann, M., Vowinckel, J., Alamo-Bethencourt, V., Wilhelm, C.S., Ahnert-Hilger, G., *et al*. (2003). Serotonylation of small GTPases is a signal transduction pathway that triggers platelet alpha-granule release. Cell *115*, 851–862.

Wandinger-Ness, A. and Zerial, M. (2014). Rab proteins and the compartmentalization of the endosomal system. Cold Spring Harb. Perspect. Biol. *6*, a022616.

Wang, H.R., Zhang, Y., Ozdamar, B., Ogunjimi, A.A., Alexandrova, E., Thomsen, G.H., and Wrana, J.L. (2003). Regulation of cell polarity and protrusion formation by targeting RhoA for degradation. Science *302*, 1775–1779.

Wei, J., Mialki, R.K., Dong, S., Khoo, A., Mallampalli, R.K., Zhao, Y., and Zhao, J. (2013). A new mechanism of RhoA ubiquitination and degradation: roles of SCF(FBXL19) E3 ligase and Erk2. Biochim. Biophys. Acta *1833*, 2757–2764.

Welch, R. A. (2016). Uropathogenic *Escherichia coli*-associated exotoxins. Microbiol. Spectr. *4*, doi:10.1128.

Wittinghofer, A. and Vetter, I. R. (2011). Structure-function relationships of the G domain, a canonical switch motif. Annu. Rev. Biochem. *80*, 943–971.

Worby, C.A., Mattoo, S., Kruger, R.P., Corbeil, L.B., Koller, A., Mendez, J.C., Zekarias, B., Lazar, C., and Dixon, J. E. (2009). The fic domain: regulation of cell signaling by adenylylation. Mol. Cell *34*, 93–103.

Xu, H., Yang, J., Gao, W., Li, L., Li, P., Zhang, L., Gong, Y. N., Peng, X., Xi, J.J., Chen, S., *et al.* (2014). Innate immune sensing of bacterial modifications of Rho GTPases by the pyrin inflammasome. Nature *513*, 237–241.

Yarbrough, M.L., Li, Y., Kinch, L.N., Grishin, N.V., Ball, H.L., and Orth, K. (2009). AMPylation of Rho GTPases by *Vibrio* VopS disrupts effector binding and downstream signaling. Science *323*, 269–272.

Yu, X., Woolery, A.R., Luong, P., Hao, Y.H., Grammel, M., Westcott, N., Park, J., Wang, J., Bian, X., Demirkan, G., *et al.* (2014). Copper-catalyzed azide-alkyne cycloaddition (click chemistry)-based detection of global pathogen-host AMPylation on self-assembled human protein microarrays. Mol. Cell Proteom. *13*, 3164–3176.

Zhang, L., Anglesio, M.S., O'Sullivan, M., Zhang, F., Yang, G., Sarao, R., Mai, P.N., Cronin, S., Hara, H., Melnyk, N., *et al.* (2007). The E3 ligase HACE1 is a critical chromosome 6q21 tumor suppressor involved in multiple cancers. Nat. Med. *13*, 1060–1069.

Zhang, L., Chen, X., Sharma, P., Moon, M., Sheftel, A.D., Dawood, F., Nghiem, M.P., Wu, J., Li, R.K., Gramolini, A.O., *et al.* (2014). HACE1-dependent protein degradation provides cardiac protection in response to haemodynamic stress. Nat. Commun. *5*, 3430.

Zhao, J., Mialki, R.K., Wei, J., Coon, T.A., Zou, C., Chen, B.B., Mallampalli, R. K., and Zhao, Y. (2013). SCF E3 ligase F-box protein complex SCF(FBXL19) regulates cell migration by mediating Rac1 ubiquitination and degradation. FASEB J. *27*, 2611–2619.

Zhao, Y. and Shao, F. (2016). Diverse mechanisms for inflammasome sensing of cytosolic bacteria and bacterial virulence. Curr. Opin. Microbiol. *29*, 37–42.

Part 3
Rho signaling in health and disease

RHOA mutations in cancer: Oncogenes or tumor suppressors? 7

Devon R. Blake and Channing J. Der*

University of North Carolina at Chapel Hill, Department of Pharmacology, Chapel Hill 27599, NC, USA

*channing_der@med.unc.edu

Keywords: RHOA, mutation, oncogene, tumor suppressor, diffuse gastric cancer, peripheral T-cell lymphoma.

7.1. Introduction

The RAS homologous (RHO) proteins comprise a major branch of the RAS super-family of small GTPases. The RHO family, which in humans contains 20 members, can be further subdivided into eight subgroups (see Chapter 1; Vega and Ridley, 2008). The most intensely studied and best understood members are RHOA, RAC1, and CDC42. RHOA and two highly related proteins, RHOB and RHOC, comprise one subgroup. With the recent identification of RHOA muta-tions in cancer, in this review, we focus on RHOA. We draw parallels with the four RAS proteins (HRAS, KRAS4A/4B, and NRAS) and the other RHO family protein found mutated in human cancers, RAC1.

7.2. RHOA structure and biochemistry

RHOA is conserved in evolution, with highly homologous and functionally related orthologs found in invertebrate species (Fig. 1). Additionally, RHOA shares 28–29% sequence identity with human RAS proteins. With two additional amino-terminal residues, the amino-acid numbering of RHOA is shifted by two. Thus, the three common hotspots for missense mutations in cancer for RAS proteins (G12, G13, and Q61) correspond to G14, G15, and Q63 in RHOA. RHOA also contains an additional 13 amino-acid "RHO insert" sequence inserted between RAS residues 123 and 124, and lacks 11 amino-acid sequences corresponding to sequences in carboxyl-terminal RAS hypervariable region (HVR) that facilitates RAS association with the inner leaflet of the plasma membrane.

Like RAS proteins, RHOA functions as a molecular switch that is active when bound to GTP and inactivated upon hydrolysis of GTP to GDP (Fig. 2). However, the intrinsic guanine nucleotide exchange and GTP hydrolysis activity of RHOA are weak and do not make for an effective regulator of signal transduction. Therefore, three classes of proteins regulate the activation state of a majority of RHO GTPases. RHOA-selective guanine nucleotide exchange factors (GEFs) stimulate RHOA exchange of GDP for GTP to form active RHOA·GTP (e.g., ECT2) (Cook *et al.*, 2013). Conversely, RHO-selective GTPase-activating proteins (GAPs) greatly accelerate RHOA GTPase activity, thereby returning RHOA to an inactive GDP-bound state (e.g., DLC1) (Yuan *et al.*, 1998). Finally, guanine nucleotide dissociation inhibitors (GDIs) sequester the inactive GDP-bound RHOA to preclude GEF activation (e.g., RHOGDI1/α) (Garcia-Mata *et al.*, 2011).

Like RAS proteins, the GDP–GTP cycle causes a change in protein conformation restricted to the Switch I (SI; a.a. 32–40) and II (SII; a.a. 62–78) regions. The active RHOA-GTP conformation binds preferentially to a spectrum of functionally diverse downstream effectors that regulate cytoplasmic signaling networks to orchestrate many cellular changes (Thumkeo *et al.*, 2013). These include actin cytoskeletal rearrangements, cell adhesion and motility, cell polarity, cell cycle progression, gene transcription, and cell growth and survival (Van Aelst and D'Souza-Schorey, 1997; Etienne-Manneville and Hall, 2002; Hall, 1998; Ridley, 2001). Key RHOA effectors include the ROCKI and ROCKII serine/threonine kinases, the citron and PKN serine/threonine kinases, mDia, Rhotekin, and Rhophilin.

Similar to RAS proteins, RHOA is synthesized initially as a cytosolic and inactive protein. A carboxyl-terminal CAAX tetrapeptide motif (C = cysteine, A = aliphatic amino acid, X = terminal amino acid) signals for covalent addition of a C20 geranylgeranyl isoprenoid lipid catalyzed by geranylgeranyltransferase-I

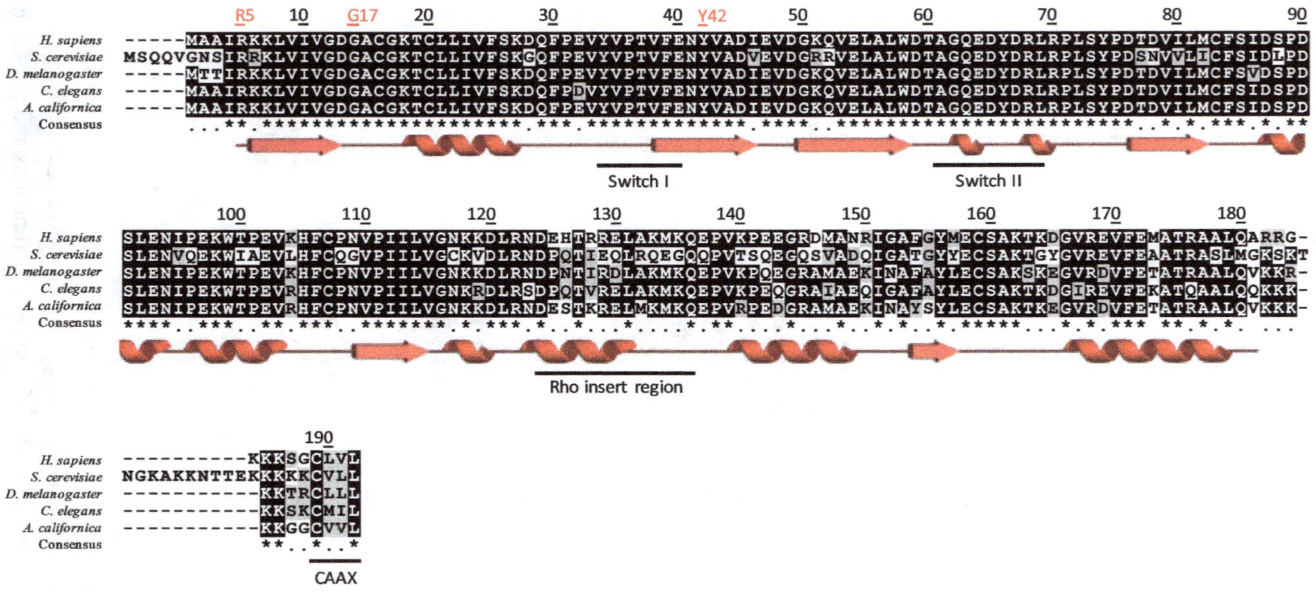

Figure 1. Conservation of RHOA protein sequence in evolution. BoxShade (SIB) was used to perform a sequence alignment of RHOA from different species, including human, yeast, fly, worm, and sea slug. The cancer-associated *RHOA* mutations, shown in red, are conserved throughout evolution. Also shown are major structural features of RHOA, including secondary structures and regions mentioned in the text.

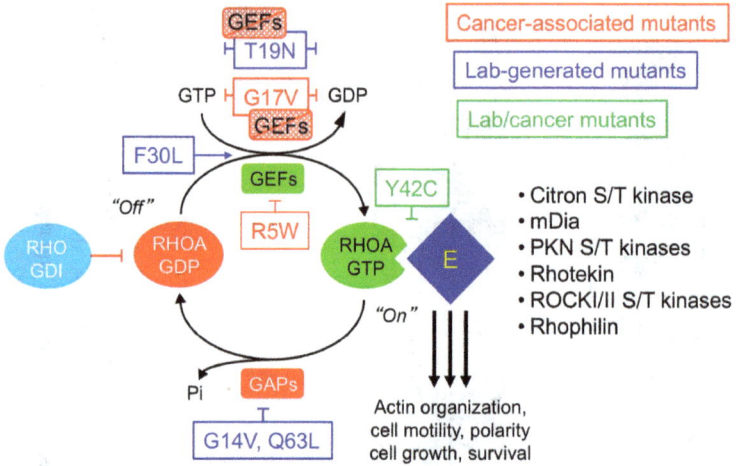

Figure 2. RHOA activation cycle and the effects of cancer-associated RHOA mutant proteins. Regulators and effectors of RHOA activity and signaling. Shown are the known and/or predicted activities of lab-generated and cancer-associated RHOA mutants. Different mutants have distinct consequences on RHOA function.

(GGT-I). This is followed by proteolytic removal of the AAX residues by RAS-converting enzyme-1 (RCE1) and methylation of the terminal geranylgeranylated cysteine residue by isocysteine methyltransferase (ICMT). Together, these modifications increase RHOA affinity for membranes. RAS proteins also undergo the same latter two CAAX-signaled modifications. However, because of different terminal X residues (M or S), RAS proteins are additionally modified by farnesyltransferase-catalyzed addition of a C15 farnesyl isoprenoid lipid. In addition to influencing GDP–GTP regulation, RHOGDIs can also regulate RHOA subcellular trafficking and membrane association by masking the geranylgeranyl group, reducing effector engagement and activation. As described above, RHOA lacks sequences corresponding to the second membrane-targeting signal present in the HVR of RAS proteins (palmitoylated cysteines and/or polylysine rich sequences) that facilitate transport to the plasma membrane. Instead, RHOA is found in Golgi and other endomembrane compartments.

RHOA forms a distinct subbranch of the RHO family together with RHOB and RHOC (83–91% sequence identity). They are most divergent at their carboxyl termini, in the region that dictates membrane targeting, as described above. Their distinct cellular properties are due in part to distinct subcellular localization. Although they share regulation by the same GEFs, GAPs, and GDIs, and interact with shared effectors, they exhibit different cellular functions and roles in human cancer (Ridley, 2013; Wheeler and Ridley, 2004).

7.3. Early studies support an oncogene function for RHOA

With strong sequence identity to RAS oncoproteins, a logical hypothesis was that RHO GTPases may also act as oncoproteins. Indeed, RHOGEFs were identified as oncogenes in the same NIH3T3 mouse fibroblast assays that identified mutant RAS in human cancers. This included DBL/MCF2 (Eva and Aaronson, 1985), VAV (Katzav et al., 1989), ECT2 (Miki et al., 1993), NET1 (Chan et al., 1996), and TIM/ARHGEF5 (Chan et al., 1994). Their activation was a result of experimentally mediated N-terminal truncations rather than bona fide cancer-associated genetic alterations; nevertheless, these findings of mechanisms that indirectly caused increased RHOA-GTP formation implied that the cancer-associated mutations in RAS that favored persistent RAS-GTP formation could similarly convert RHOA into an oncogene.

To evaluate this possibility, lab-generated RHOA variants with single-amino-acid substitutions analogous to those found in RAS (G12V and Q61L) were generated in RHOA (G14V and Q63L) and exhibited gain-of-function properties. These activated mutants were used as key reagents to implicate RHOA in cell surface receptor-mediated activation of RHOA and promotion of actin stress fibers and focal adhesions. However, when these lab-generated RHOA mutants were evaluated in NIH 3T3 fibroblast growth transformation assays, they did not drive the same robust growth transformation seen with mutant RAS proteins (Khosravi-Far et al., 1995; Lin et al., 1999; Qiu et al., 1995; Zohn et al., 1998). However, they did cooperate with weakly activated mutants of the CRAF/RAF1 serine/threonine kinase and caused potent focus formation. Furthermore, RHOA function was shown to be required for HRAS transformation of rodent fibroblasts (Khosravi-Far et al., 1995; Qiu et al., 1995).

Further support for a RHOA oncogenic function came from findings that RHOGEFs act as drivers in cancer. In particular, ECT2, a RHOA (and RHOB and RHOC) selective GEF, is overexpressed in lung, ovarian, and other cancers, and suppression of ECT2 impaired tumor growth *in vitro* and *in vivo* (Huff et al., 2013). Finally, similar to studies in NIH3T3 cells, ECT2 function was found to be required for *KRAS*-driven lung cancer growth (Justilien et al., 2017).

RHOGAP activity is lost in some cancers, resulting in indirect activation of RHOA. In particular, DLC1 is a RHOA-selective GAP that was identified initially as deleted in lung cancer (Yuan et al., 1998). Subsequent studies found promoter methylation and loss of DLC1 expression in many cancer types (Yuan et al., 2003). Furthermore, DLC1 loss can also involve an E3 ligase-directed mechanism driving DLC1 degradation in lung cancer (Kim et al., 2013).

Despite the substantial experimental evidence that RHOA can function as an oncogene, exome sequencing studies of major genetic alterations in common cancers failed to identify missense mutations analogous to those found in RAS (Agrawal *et al.*, 2011; Parsons *et al.*, 2010; Sjöblom *et al.*, 2006; Wood *et al.*, 2007). However, in 2012, exome sequencing studies identified missense mutations in RAC1 in sun-exposed melanomas (Hodis *et al.*, 2012; Krauthammer *et al.*, 2012). Unexpectedly, these mutations were found at RAC1 residue P29 rather than at the hotspots found in RAS: G12, G13, or Q61. However, the P29S mutation causes a "fast cycling" defect, with a three-fold enhanced rate of intrinsic nucleotide exchange and no defect in GTP hydrolysis (Davis *et al.*, 2013). Since intracellular GTP levels are 10-fold higher than GDP, this defect favors increased formation of active RAC1-GTP. Although the G12, G13, and Q61 mutations in RAS also result in increased RAS-GTP formation, this occurs primarily due to impaired intrinsic and GAP-stimulated GTP hydrolysis activity.

Why the classical RAS GTP hydrolysis-deficient mutations are not found in RAC1 in cancer is not clear. One possible explanation is provided by the observation that a lab-generated fast-cycling mutant of CDC42, F28L, caused growth transformation of NIH 3T3 cells (Lin *et al.*, 1997), whereas GTPase-deficient CDC42 mutants were growth inhibitory. Since a F28L mutant of HRAS shows a weaker transforming activity than G12V (Reinstein *et al.*, 1991), perhaps GTPase-deficient mutants of RAC1 are too potently activated and not well tolerated in cancer. Alternatively, whereas a GTPase deficiency results in a chronically GTP-bound protein with persistent association with effectors, the fast GDP–GTP cycling of RAC1 may more effectively activate effector function.

7.4. *RHOA* mutations discovered in cancer

In 2014, three independent groups uncovered recurrent *RHOA* mutations in diffuse gastric cancer, and very surprisingly, not at the classical mutational hotspots found in RAS or the fast-cycling mutation found in RAC1 (Fig. 3) (Bass *et al.*, 2014; Kakiuchi *et al.*, 2014; Wang *et al.*, 2014). Instead, the three main hotspots for *RHOA* mutations in diffuse gastric cancer are R5, G17, and Y42 (Figs. 3 and 4). That same year, three groups published their discoveries of *RHOA* mutations in specific subtypes of peripheral T-cell lymphomas (Palomero *et al.*, 2014; Sakata-Yanagimoto *et al.*, 2014; Yoo *et al.*, 2014). These include angioimmunoblastic T-cell lymphoma (AITL) and peripheral T-cell lymphoma not otherwise specified (PTCL-NOS). Intriguingly, over 90% of *RHOA* mutations in these lymphomas are found at G17, which is located in the nucleotide-binding pocket. Importantly, none of the cancer-associated *RHOA* mutations are homologous to activating *RAS* or *RAC1* mutations (Fig. 3(b)). As summarized below, the limited information on

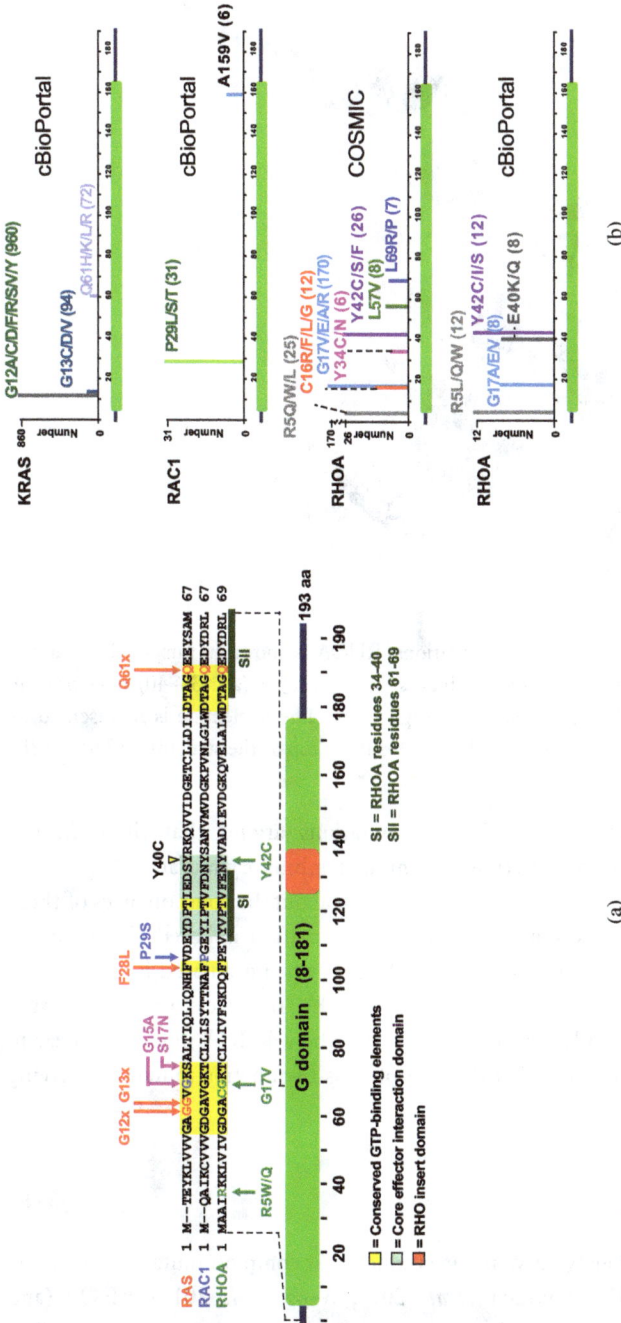

(a)

(b)

Figure 3. (a) Comparison of mutations in small GTPases. Cancer-associated RHOA missense mutations are located at different residues when compared with the gain-of-function activating mutations found in RAS and RAC1. Clustal/W was used to align the N-terminal sequences of RAS, RAC1, and RHOA. The proximity of the RHOA mutants to key RHOA sequence elements important for GDP–GTP cycling and/or interaction with GEFs, GAPs, and/or effectors are indicated. (b) Cancer-associated mutations in KRAS and RHO GTPases. Data were compiled from either cBioPortal or COSMIC. Only residues with five or more missense mutations are shown. The different amino acid substitutions at each hotspot are indicated in the order of decreasing frequency. *RHOA* mutations are at different codons than the activating mutations found in KRAS and RAC1.

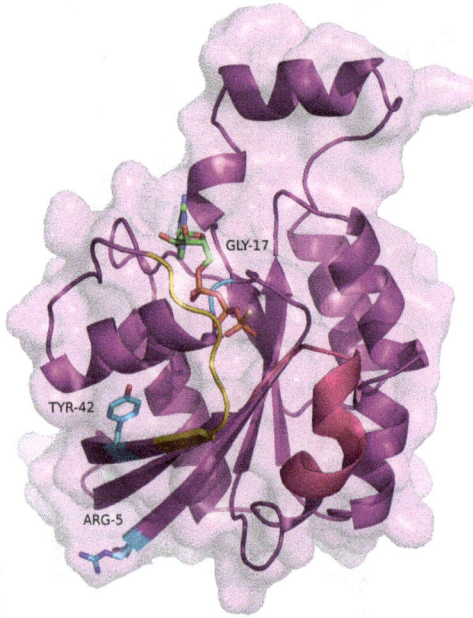

Figure 4.　Cancer-associated *RHOA* mutations. RHOA is shown in purple. The cancer-associated mutations are represented by blue sticks. Switch I (a.a. 34–40) is colored in yellow. Switch II (a.a. 61–69) is colored in bright pink. The nucleotide is represented by green sticks. The RHO insert region is the α helix at the top of the structure. PDB: 1A2B.

the biochemical consequences of these mutations argues that these *RHOA* mutations are either loss-of-function or dominant-inhibitory mutants.

　　Below we summarize what is currently known about the consequences of these mutations on RHOA function. The current information is based either on the consequences of analogous mutations in HRAS or when evaluated in the context of an activated GTPase-deficient mutant of RHOA. It is clear that the current information fails to provide a clear basis for why these RHOA mutants appear to act more as gain-of-function rather than simple loss-of-function mutants in driving cancer growth.

7.4.1. *Arginine 5*

Mutation of R5 to either Q or W is the second most common mutation in gastric cancer (Bass *et al.*, 2014; Kakiuchi *et al.*, 2014; Wang *et al.*, 2014). RHOA (and RHOB and RHOC) contains a unique arginine at position 5 not conserved in other

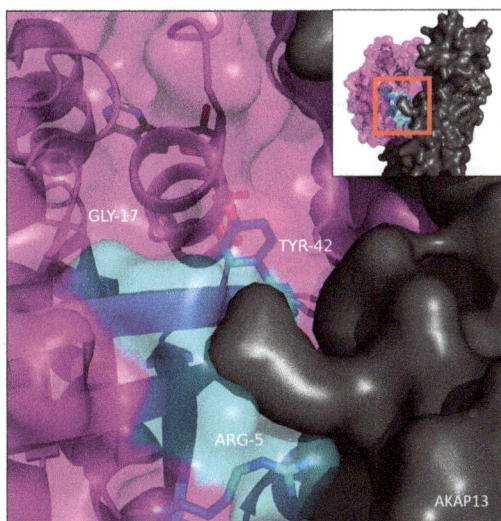

Figure 5. R5 and Y42 found in and around the RHOGEF "specificity patch". Snyder *et al.* (2002) determined the basis for selective activation of RHOA and not RAC1 or CDC42 by RHOA (and RHOB and RHOC) specific GEFs. RHOA is shown in purple, and the "specificity patch" is shown in teal. AKAP13 RHOGEF is shown in gray. The specificity patch contains residues on RHOA that form productive interactions with GEFs. R5 is one such residue that forms a polar interaction with an E residue on GEFs. Although Y42 does not make direct contact with GEFs, it is directly adjacent to the "specificity patch" (which includes V43), and the Y42C mutation could alter the GEF interaction surface. PDB: 4D0N.

RHO family small GTPases (RAC1 and CDC42) or in RAS. Conserved in evolution and found in invertebrate orthologs of RHOA, this residue was shown to be important for GEF recognition of RHOA (Derewenda *et al.*, 2004; Snyder *et al.*, 2002). R5 forms a polar interaction with Q752 of the RHOGEF DBS/MCF2L, an interaction which does not occur between DBS/MCF2L and either RAC1 or CDC42. Thus, it was found that R5 is part of a "specificity patch" within RHOA that mediates the interaction between the DBL family of RHOGEFs and their substrates (Fig. 5) (Snyder *et al.*, 2002). Consistent with these data, the crystal structure of RHOA bound to the DBL family member PDZRhoGEF/AKAP13 showed formation of a salt bridge between R5 of RHOA and D873 of PDZRhoGEF/AKAP13 (Derewenda *et al.*, 2004). Moreover, several other RhoGEFs contain D and E amino acids at analogous residues, indicating that this interaction plays an important role for RHOGEF recognition of RHOA. Consequently, mutation of the R5 residue could affect GEF-mediated nucleotide exchange, although this hypothesis has not been tested.

7.4.2. *Glycine 17*

The two predominant mutations found at G17 are G17V and G17E, with G17V being the far more prevalent substitution (Bass *et al.*, 2014; Kakiuchi *et al.*, 2014; Palomero *et al.*, 2014; Sakata-Yanagimoto *et al.*, 2014; Wang *et al.*, 2014; Yoo *et al.*, 2014). G17 is the predominant mutation in the peripheral T-cell lymphomas AITL and PTCL-NOS and is found in 65% of AITL (53.3% or 24/45 (Yoo *et al.*, 2014), 67% or 22/35 (Palomero *et al.*, 2014), 70.8% or 51/72 (Sakata-Yanagimoto *et al.*, 2014)) and 15% of PTCL-NOS (7.7% or 1/13 (Yoo *et al.*, 2014), 17.2% or 15/87 (Sakata-Yanagimoto *et al.*, 2014); 8/44 or 18% (Palomero *et al.*, 2014)) cases. It is found within the nucleotide-binding pocket, and its backbone amide hydrogen binds to both the γ-phosphate and the oxygen in the first phosphodiester bond of guanine nucleotide (Fig. 6). The smallness of its side chain allows space for the guanine base to fit within the nucleotide-binding pocket. Structural analysis of the RHOA protein reveals that mutation of G17 to any other amino acid would occlude the guanosine base in the nucleotide-binding pocket and thus the protein would be

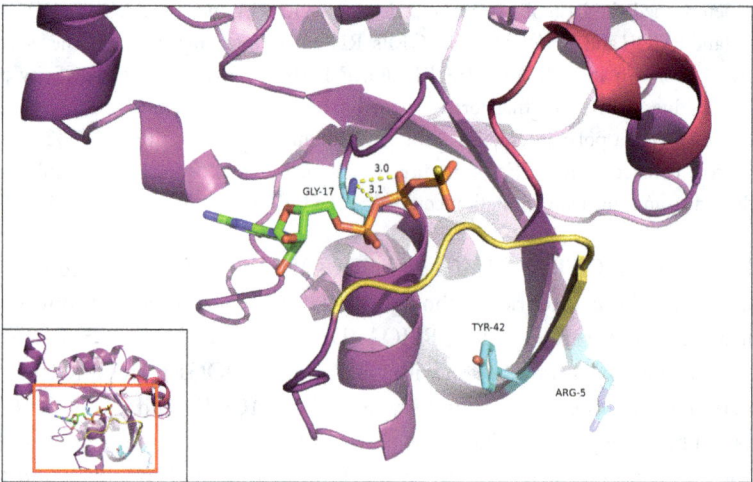

Figure 6. G17 is found in the nucleotide-binding pocket. RHOA is shown in purple. The cancer-associated mutations are represented in blue sticks. Switch I (a.a. 34–40) is colored in yellow. Switch II (a.a. 61–69) is colored in bright pink. The nucleotide is represented in green sticks. The backbone amide nitrogen forms two hydrogen bonds with the nucleotide, the first with the first phosphodiester bond and the second with the β phosphate. There is no room for a side chain from amino acid 17 in the nucleotide-binding pocket. G17 mutations preclude nucleotide binding due to steric occlusion of the side chain and the guanine base of the nucleotide. PDB: 1A2B.

found in a nucleotide-free state. Because nucleotide-free GTPases have higher affinity for GEFs, they are speculated to acquire a dominant-negative function whereby they form nonproductive complexes with GEFs. Impairment of RHOA-selective GEFs in this manner by mutant RHOA in turn prevents activation of wild-type RHOA.

Evidence for this function is based on studies with analogous mutants of HRAS. The original premise of an HRAS dominant negative came from the identification of the HRAS S17N mutant that was deficient in GTP but not GDP binding (this was later corrected) that caused growth suppression of NIH3T3 cells by blocking endogenous RAS activation (Farnsworth and Feig, 1991; Stacey et al., 1991). Independently, a functional screen in S. cerevisiae identified mutants of yeast RAS that acted as inhibitors of WT yeast RAS function (Powers et al., 1989). One mutation generated in the analogous amino acid in HRAS, G15A, was shown to act as a dominant negative, nucleotide-free mutant with potent growth-suppressing activity (Chen et al., 1994). Subsequent studies introduced this mutant into RHO GTPases, and the mutants were utilized as affinity reagents to isolate interacting GEFs (Arthur et al., 2002; Reuther et al., 2001). Since RHOA-selective GEFs commonly also activate RHOB and RHOC, such mutants impair the activity of multiple RHO GTPases. Furthermore, a subset of RhoGEFs are broadly active on multiple RHO GTPases (e.g., VAV activates RHOA, RAC1, CDC42, and RHOG), the *RHOA* G17 mutations likely have broader implications beyond blocking RHOA activation.

Although they are still not well understood, RHOA G17 mutants are the best characterized of all cancer-associated RHOA mutants. Consistent with the hypothesis of these mutants functioning as nucleotide-free dominant negatives, purified RHOA G17V showed severely impaired nucleotide binding and exchange as measured by [^{35}S]GTPγS binding (Nagata et al., 2016) and mant-GTP loading (Palomero et al., 2014). Furthermore, RHOA G17V expressed exogenously in NIH3T3 cells showed increased binding to the RHOA-selective GEF ECT2 as compared to RHOA WT, attenuated the activity of a SRF-RE transcription reporter, and inhibited formation of stress fibers (Sakata-Yanagimoto et al., 2014). All these results are consistent with a dominant negative function. Several groups also showed that exogenous expression of RhoA G17 mutants in NIH3T3 cells (Sakata-Yanagimoto et al., 2014) and Jurkat cells (Palomero et al., 2014) attenuated total cellular RHOA-GTP levels. Interestingly, ectopic expression of either a RHOA G17V or T19N mutant in Jurkat, SUP-T1, or MOLT-4 cells — all of T-cell lineage — caused cells to grow faster than ectopic expression of either WT or the activated G14V RHOA. Additionally, exogenous expression of either RHOA G17V or T19N also increased the invasive potential of Jurkat and H9 cells

(Yoo *et al.*, 2014). These data indicate that attenuation of RHOA activation caused increased proliferation and invasive potential of T cells. Knockdown of *RHOA* in the G17E-mutant breast cancer cell line BT474 and colon cancer cell line SW948 decreased their ability to form 3D spheroids, and this growth defect could not be rescued by re-expression of WT RHOA (Kakiuchi *et al.*, 2014). Taken together, these data strongly suggest that G17 RHOA mutants function as dominant negatives to attenuate WT RHOA signaling, although how this phenomenon drives cellular proliferation and invasion remains to be elucidated.

7.4.3. Tyrosine 42

RHOA Y42C is the most common *RHOA* mutation in diffuse gastric cancers (Bass *et al.*, 2014; Kakiuchi *et al.*, 2014; Wang *et al.*, 2014). The analogous mutation in HRAS, Y40C, was first identified in a random mutagenesis yeast two-hybrid binding assay to identify activated HRAS mutants with impaired binding to its effector, CRAF. Researchers trying to parse out the role of downstream signaling pathways in RAS-induced transformation generated artificial RAS mutants that exhibited impaired binding to effectors. Among these was HRAS G12V/Y40C, which showed impaired binding to CRAF/RAF1 and RALGEF effectors (Khosravi-Far *et al.*, 1996; Rodriguez-Viciana *et al.*, 1997) but retained binding to phosphoinositide-3 kinase (PI3K) (Rodriguez-Viciana *et al.*, 1997). This mutant also retained the ability to synergize with a weakly activated CRAF/RAF-1 mutant to transform NIH3T3 cells, indicating that it still retained the ability to activate a subset of downstream effectors that drive RAS-dependent growth transformation (Khosravi-Far *et al.*, 1996; Rodriguez-Viciana *et al.*, 1997).

Having found that "effector binding" mutants could inform understanding of RAS biology, researchers then generated the analogous mutants in RHO family members to study RHO biology. For example, a RAC1 Y40C/Q61L mutant was unable to bind the PAK serine/threonine kinase but was able to activate ROCK and induce formation of lamellipodia (Joneson *et al.*, 1996; Lamarche *et al.*, 1996; Westwick *et al.*, 1997). Similarly, a CDC42 Y40C/Q61L mutant did not bind PAK or other CRIB domain-containing proteins that were tested but was still able to induce formation of filopodia (Lamarche *et al.*, 1996). Comparable experiments were also performed on RHOA. An RHOA G14V/Y42C mutant was found to be deficient in PKN binding but retained binding in all other effectors. Consistent with these results, this mutant was able to induce formation of stress fibers in NIH 3T3 cells (Sahai *et al.*, 1998). Interestingly, an RhoA Y42C/Q63L mutant was unable to activate the SRF transcription factor, lost its ability to transform NIH3T3 cells, and lost its ability to form tumors in a flank xenograft model

(Zohar *et al.*, 1998). The fact that there are two conflicting results with two different activated mutants (i.e., Y42C in the context of either G14V or Q63L) highlights the need for studies of the Y42C mutation in the context of RHOA WT.

There has been limited analysis of the biochemical and cellular effects of the RHOA Y42C mutation in the context of RHOA WT. Exogenous expression of RHOA Y42C decreased total RHOA-GTP levels by about 50% in HEK293T/17 cells, as measured by a Rhotekin-RBD pulldown assay (Wang *et al.*, 2014). One caution with these analyses is that it is not known if the Y42C mutation impairs Rhotekin-RBD binding, which would negate this assay as an accurate measure of the level of GTP-bound RHOA. Moreover, exogenous expression of RHOA Y42C but not RHOA WT enabled intestinal organoids to grow in the absence of the ROCK inhibitor Y-27632, suggesting a possible defect in ROCK signaling by Y42C (Wang *et al.*, 2014). Lastly, *RHOA* knockdown in OE19 cells, an esophageal carcinoma cell line harboring a *RHOA* Y42C mutation, inhibited growth in a 3D spheroid-formation assay (Kakiuchi *et al.*, 2014). However, all these experiments were performed in nondiffuse gastric cancer systems, and more detailed analyses in more relevant cancer cell model systems are needed to fully elucidate the role of Y42C in diffuse gastric cancer progression.

7.5. Conclusions and future insights

Although many initial studies proposed an oncogenic role for RHOA, early exome sequencing studies of common cancers failed to identify *RHOA* mutations in cancer (Agrawal *et al.*, 2011; Parsons *et al.*, 2010; Sjöblom *et al.*, 2006; Wood *et al.*, 2007). Recent sequencing studies, however, have now identified recurrent *RHOA* mutations in diffuse gastric cancer and in the peripheral T-cell lymphomas AITL and PTCL-NOS (Bass *et al.*, 2014; Kakiuchi *et al.*, 2014; Palomero *et al.*, 2014; Sakata-Yanagimoto *et al.*, 2014; Wang *et al.*, 2014; Yoo *et al.*, 2014). In contrast to *RAS* mutations, which can be found in a diverse spectrum of cancer types, the spectrum of cancers with *RHOA* mutations is quite narrow and limited. While the reason for such a narrow cell type distribution is not understood, it may reflect the possibility that these aberrant *RHOA* mutations can act as drivers in only specific tissues or genetic backgrounds. Intriguingly, none of these mutations are at the classical mutational hotspots found in *RAS*, and their functions are largely unknown. Moreover, the fact that many of these mutations are heterozygous and therefore exert their effects in the presence of the wild-type allele suggests that they function either as dominant negative or dominant activating mutations and not as simple loss-of-function mutations. The best characterized of these mutations, G17V, is a putative dominant-negative mutation and acts by attenuating

RHO signaling (Nagata *et al.*, 2016; Palomero *et al.*, 2014; Sakata-Yanagimoto *et al.*, 2014; Yoo *et al.*, 2014), but the mechanism by which decreased RHO signaling enhances tumor growth and proliferation remains to be elucidated. The most frequent mutation in diffuse gastric cancer and second-most frequent mutation agnostic of cancer type is Y42C. Preliminary studies indicate that this mutation alters the interactions between RHOA and its effectors (Sahai *et al.*, 1998; Zohar *et al.*, 1998) and also attenuates RHOA·GTP levels in cells (Wang *et al.*, 2014). However, no mechanistic studies have been performed to determine how this mutation drives diffuse gastric cancer growth. The third-most common *RHOA* mutations in diffuse gastric cancer, R5Q or R5W, have not been studied at all.

Over 90% of *RHOA* mutations in AITL and PTCL-NOS are G17V, whereas the spectrum of *RHOA* mutations in diffuse gastric cancer is more diverse. It is apparent that although the net cellular effect of the different *RHOA* mutations may be the same (tumor growth and progression), each mutation likely acts through distinct mechanisms to achieve the same cellular effect. Additional studies are needed to dissect the distinct mechanisms of different RHOA mutants.

In summary, the unique mutational spectrum found in *RHOA* in cancer clearly demonstrates the that aberrant RHOA function can drive cancer development and growth. However, only with further detailed structural, biochemical, and cellular analyses will the mechanistic basis for how mutant RHOA drives cancer be fully understood. Only when this is understood can approaches for pharmacologic intervention for cancers with *RHOA* mutations be identified and pursued.

References

Van Aelst, L. and D'Souza-Schorey, C. (1997). Rho GTPases and signaling networks. Genes Dev. *11*, 2295–2322.

Agrawal, N., Frederick, M.J., Pickering, C.R., Bettegowda, C., Chang, K., Li, R.J., Fakhry, C., Xie, T., Zhang, J., Wang, J., *et al.* (2011). Exome sequencing of head and neck squamous cell carcinoma reveals inactivating mutations in NOTCH1. Science *333*, 1154–1157.

Arthur, W.T., Ellerbroek, S.M., Der, C.J., Burridge, K., and Wennerberg, K. (2002). XPLN, a guanine nucleotide exchange factor for RhoA and RhoB, but not RhoC. J. Biol. Chem. *277*, 42964–42972.

Bass, A.J., Thorsson, V., Shmulevich, I., Reynolds, S.M., Miller, M., Bernard, B., Hinoue, T., Laird, P.W., Curtis, C., Shen, H., *et al.* (2014). Comprehensive molecular characterization of gastric adenocarcinoma. Nature *513*, 202–209.

Chan, A.M., McGovern, E.S., Catalano, G., Fleming, T.P., and Miki, T. (1994). Expression cDNA cloning of a novel oncogene with sequence similarity to regulators of small GTP-binding proteins. Oncogene *9*, 1057–1063.

Chan, A.M., Takai, S., Yamada, K., and Miki, T. (1996). Isolation of a novel oncogene, NET1, from neuroepithelioma cells by expression cDNA cloning. Oncogene *12*, 1259–1266.

Chen, S.Y., Huff, S.Y., Lai, C.C., Der, C.J., and Powers, S. (1994). Ras-15A protein shares highly similar dominant-negative biological properties with Ras-17N and forms a stable, guanine-nucleotide resistant complex with CDC25 exchange factor. Oncogene *9*, 2691–2698.

Cook, D.R., Rossman, K.L., and Der, C.J. (2013). Rho guanine nucleotide exchange factors: regulators of Rho GTPase activity in development and disease. Oncogene *33*, 4021–4035.

Davis, M.J., Hak, B., Holman, E.C., Halaban, R., Schlessinger, J., Boggon, T.J., and Ha, B.H. (2013). RAC1P29S is a spontaneously activating cancer-associated GTPase. Proc. Natl. Acad. Sci. USA *110*, 912–917.

Derewenda, U., Oleksy, A., Stevenson, A.S., Korczynska, J., Dauter, Z., Somlyo, A.P., Otlewski, J., Somlyo, A.V., and Derewenda, Z.S. (2004). The crystal structure of RhoA in complex with the DH/PH fragment of PDZRhoGEF, an activator of the Ca2+ sensitization pathway in smooth muscle. Structure *12*, 1955–1965.

Etienne-Manneville, S. and Hall, A. (2002). Rho GTPases in cell biology. Nature *420*, 629–635.

Eva, A. and Aaronson, S.A. (1985). Isolation of a new human oncogene from a diffuse B-cell lymphoma. Nature *316*, 273–275.

Farnsworth, C.L. and Feig, L.A. (1991). Dominant inhibitory mutations in the Mg(2+)-binding site of RasH prevent its activation by GTP. Mol. Cell. Biol. *11*, 4822–4829.

Garcia-Mata, R., Boulter, E., and Burridge, K. (2011). The "invisible hand": regulation of RHO GTPases by RHOGDIs. Nat. Rev. Mol. Cell Biol. *12*, 493–504.

Hall, A. (1998). Rho GTPases and the actin cytoskeleton. Science *279*, 509–514.

Hodis, E., Watson, I.R., Kryukov, G.V, Arold, S.T., Imielinski, M., Theurillat, J.P., Nickerson, E., Auclair, D., Li, L., Place, C., et al. (2012). A landscape of driver mutations in melanoma. Cell *150*, 251–263.

Huff, L.P., Decristo, M.J., Trembath, D., Kuan, P.F., Yim, M., Liu, J., Cook, D.R., Miller, C.R., Der, C.J., and Cox, A.D. (2013). The role of Ect2 nuclear RhoGEF activity in ovarian cancer cell transformation. Genes Cancer *4*, 460–475.

Joneson, T., McDonough, M., Bar-Sagi, D., and Van Aelst, L. (1996). RAC regulation of actin polymerization and proliferation by a pathway distinct from Jun kinase. Science *274*, 1374–1376.

Justilien, V., Ali, S.A., Jamieson, L., Yin, N., Cox, A.D., Der, C.J., Murray, N.R., Fields, A.P. (2017). Ect2-dependent rRNA synthesis is required for KRAS-TRP53-driven lung adenocarcinoma. Cancer Cell *31*, 367–378.

Kakiuchi, M., Nishizawa, T., Ueda, H., Gotoh, K., Tanaka, A., Hayashi, A., Yamamoto, S., Tatsuno, K., Katoh, H., Watanabe, Y., et al. (2014). Recurrent gain-of-function mutations of RHOA in diffuse-type gastric carcinoma. Nat. Genet. *46*, 583–587.

Katzav, S., Martin-Zanca, D., and Barbacid, M. (1989). Vav, a novel human oncogene derived from a locus ubiquitously expressed in hematopoietic cells. EMBO J. *8*, 2283–2290.

Khosravi-Far, R., Solski, P.A., Clark, G.J., Kinch, M.S., and Der, C.J. (1995). Activation of Rac1, RhoA, and mitogen-activated protein kinases is required for Ras transformation. Mol. Cell. Biol. *15*, 6443–6453.

Khosravi-Far, R., White, M.A., Westwick, J.K., Solski, P.A., Chrzanowska-Wodnicka, M., Van Aelst, L., Wigler, M.H., and Der, C.J. (1996). Oncogenic Ras activation of Raf/ mitogen-activated protein kinase-independent pathways is sufficient to cause tumorigenic transformation. Mol. Cell. Biol. *16*, 3923–3933.

Kim, T.Y., Jackson, S., Xiong, Y., Whitsett, T.G., Lobello, J.R., Weiss, G.J., Le Tran, N., Bang, Y.-J., and Der, C.J. (2013). CRL4A-FBXW5-mediated degradation of DLC1 Rho GTPase-activating protein tumor suppressor promotes non-small cell lung cancer cell growth. Proc. Natl. Acad. Sci. USA *110*, 16868–16873.

Krauthammer, M., Kong, Y., Ha, B.H., Evans, P., Bacchiocchi, A., McCusker, J.P., Cheng, E., Davis, M.J., Goh, G., Choi, M., et al. (2012). Exome sequencing identifies recurrent somatic RAC1 mutations in melanoma. Nat. Genet. *44*, 1006–1014.

Lamarche, N., Tapon, N., Stowers, L., Burbelo, P.D., Aspenström, P., Bridges, T., Chant, J., and Hall, A. (1996). Rac and Cdc42 induce actin polymerization and G1 cell cycle progression independently of p65(PAK) and the JNK/SAPK MAP kinase cascade. Cell *87*, 519–529.

Lin, R., Bagrodia, S., Cerione, R., and Manor, D. (1997). A novel Cdc42Hs mutant induces cellular transformation. Curr. Biol. *7*, 794–797.

Lin, R., Cerione, R.A., and Manor, D. (1999). Specific contributions of the small GTPases Rho, Rac, and Cdc42 to Dbl transformation. J. Biol. Chem. *274*, 23633–23641.

Miki, T., Smith, C.L., Long, J.E., Eva, A., and Fleming, T.P. (1993). Oncogene ect2 is related to regulators of small GTP-binding proteins. Nature *362*, 462–465.

Nagata, Y., Kontani, K., Enami, T., Kataoka, K., Ishii, R., Totoki, Y., Kataoka, T.R., Hirata, M., Aoki, K., Nakano, K., et al. (2016). Variegated RHOA mutations in adult T-cell leukemia/lymphoma. Blood *127*, 596–604.

Palomero, T., Couronne, L., Khiabanian, H., Kim, M.Y., Ambesi-Impiombato, A., Perez-Garcia, A., Carpenter, Z., Abate, F., Allegretta, M., Haydu, J.E., et al. (2014). Recurrent mutations in epigenetic regulators, RHOA and FYN kinase in peripheral T cell lymphomas. Nat. Genet. *46*, 166–170.

Parsons, D.W., Jones, S., Zhang, X., Lin, J.C., Leary, R.J., Angenendt, P., Mankoo, P., Carter, H., Siu, I., et al. (2010). An integrated genomic analysis of human glioblastoma multiforme. Science *1807*, 1807–1813.

Powers, S., O'Neill, K., and Wigler, M. (1989). Dominant yeast and mammalian RAS mutants that interfere with the CDC25-dependent activation of wild-type RAS in *Saccharomyces cerevisiae*. Mol. Cell. Biol. *9*, 390–395.

Qiu, R.G., Chen, J., McCormick, F., and Symons, M. (1995). A role for Rho in Ras transformation. Proc. Natl. Acad. Sci. USA *92*, 11781–11785.

Reinstein, J., Schlichting, I., Frech, M., Goody, R.S., and Wittinghofer, A. (1991). p21 with a phenylalanine 28 — leucine mutation reacts normally with the GTPase activating protein GAP but nevertheless has transforming properties. J. Biol. Chem. *266*, 17700–17706.

Reuther, G.W., Lambert, Q.T., Booden, M.A., Wennerberg, K., Becknell, B., Marcucci, G., Sondek, J., Caligiuri, M.A., and Der, C.J. (2001). Leukemia-associated Rho guanine nucleotide exchange factor, a Dbl family protein found mutated in leukemia, causes transformation by activation of RhoA. J. Biol. Chem. *276*, 27145–27151.

Ridley, A.J. (2001). Rho family proteins: coordinating cell responses. Trends Cell Biol. *11*, 471–477.

Ridley, A.J. (2013). RhoA, RhoB and RhoC have different roles in cancer cell migration. J. Microsc. *251*, 242–249.

Rodriguez-Viciana, P., Warne, P.H., Khwaja, A., Marte, B.M., Pappin, D., Das, P., Waterfield, M.D., Ridley, A., and Downward, J. (1997). Role of phosphoinositide 3-OH kinase in cell transformation and control of the actin cytoskeleton by Ras. Cell *89*, 457–467.

Sahai, E., Alberts, A.S., and Treisman, R. (1998). RhoA effector mutants reveal distinct effector pathways for cytoskeletal reorganization, SRF activation and transformation. EMBO J. *17*, 1350–1361.

Sakata-Yanagimoto, M., Enami, T., Yoshida, K., Shiraishi, Y., Ishii, R., Miyake, Y., Muto, H., Tsuyama, N., Sato-Otsubo, A., Okuno, Y., *et al.* (2014). Somatic RHOA mutation in angioimmunoblastic T cell lymphoma. Nat. Genet. *46*, 171–175.

Sjöblom, T., Jones, S., Wood, L.D., Parsons, D.W., Lin, J., Barber, T.D., Mandelker, D., Leary, R.J., Ptak, J., Silliman, N., *et al.* (2006). The consensus coding sequences of human breast and colorectal cancers. Science *314*, 268–274.

Snyder, J.T., Worthylake, D.K., Rossman, K.L., Betts, L., Pruitt, W.M., Siderovski, D.P., Der, C.J., and Sondek, J. (2002). Structural basis for the selective activation of Rho GTPases by Dbl exchange factors. Nat. Struct. Biol. *9*, 468–475.

Stacey, D.W., Feig, L.A., and Gibbs, J.B. (1991). Dominant inhibitory Ras mutants selectively inhibit the activity of either cellular or oncogenic Ras. Mol. Cell. Biol. *11*, 4053–4064.

Thumkeo, D., Watanabe, S., and Narumiya, S. (2013). Physiological roles of rho and rho effectors in mammals. Eur. J. Cell Biol. *92*, 303–315.

Vega, F.M. and Ridley, A.J. (2008). Rho GTPases in cancer cell biology. FEBS Lett. *582*, 2093–2101.

Wang, K., Yuen, S.T., Xu, J., Lee, S.P., Yan, H.H.N., Shi, S.T., Siu, H.C., Deng, S., Chu, K.M., Law, S., *et al.* (2014). Whole-genome sequencing and comprehensive molecular profiling identify new driver mutations in gastric cancer. Nat. Genet. *46*, 573–582.

Westwick, J.K., Lambert, Q.T., Clark, G.J., Symons, M., Van Aelst, L., Pestell, R.G., and Der, C.J. (1997). Rac regulation of transformation, gene expression, and actin organization by multiple, PAK-independent pathways. Mol. Cell. Biol. *17*, 1324–1335.

Wheeler, A.P. and Ridley, A.J. (2004). Why three Rho proteins? RhoA, RhoB, RhoC, and cell motility. Exp. Cell Res. *301*, 43–49.

Wood, L.D., Parsons, D.W., Jones, S., Lin, J., Sjoblom, T., Leary, R.J., Shen, D., Boca, S.M., Barber, T., Ptak, J., *et al.* (2007). The genomic landscapes of human breast and colorectal cancers. Science *318*, 1108–1113.

Yoo, H.Y., Sung, M.K., Lee, S.H., Kim, S., Lee, H., Park, S., Kim, S.C., Lee, B., Rho, K., Lee, J.E., *et al.* (2014). A recurrent inactivating mutation in RHOA GTPase in angio-immunoblastic T cell lymphoma. Nat. Genet. *46*, 371–375.

Yuan, B., Miller, M.J., Keck, C.L., Zimonjic, D.B., Thorgeirsson, S.S., and Popescu, N.C. (1998). Cloning, characterization, and chromosomal localization of a gene frequently deleted in human liver cancer (DLC-1) homologous to rat RhoGAP. Cancer Res. *58*, 2196–2199.

Yuan, B.Z., Durkin, M.E., and Popescu, N.C. (2003). Promoter hypermethylation of DLC-1, a candidate tumor suppressor gene, in several common human cancers. Cancer Genet. Cytogenet. *140*, 113–117.

Zohar, M., Teramoto, H., Katz, B.Z., Yamada, K.M., and Gutkind, J.S. (1998). Effector domain mutants of Rho dissociate cytoskeletal changes from nuclear signaling and cellular transformation. Oncogene *17*, 991–998.

Zohn, I.M., Campbell, S.L., Khosravi-Far, R., Rossman, K.L., and Der, C.J. (1998). Rho family proteins and Ras transformation: the RHOad less traveled gets congested. Oncogene *17*, 1415–1438.

Modulation of osteoclast differentiation and function by Rho GTPases

8

Yongqiang Wang, Patricia Joyce Brooks, and Michael Glogauer*

Matrix Dynamics Group, Faculty of Dentistry, University of Toronto, Toronto, Ontario M5S 3E2, Canada

*Michael.Glogauer@utoronto.ca

Keywords: Rho GTPases, reactive oxygen species, actin filaments, podosome, fusopods, osteoclast differentiation, osteoclast function.

8.1. Bone and bone cells

Bone is a hard connective tissue that provides structure, support, and protection, as well as a source of minerals and cells. It is a dynamic structure, constantly remodeling in order to respond to mechanical loading and growth factors and undergoes repair and provides an environment for the maturation of bone marrow cells.

The four major resident bone cells are osteoblasts (OBs), bone lining/endosteal cells, osteocytes, and osteoclasts (OCs). OBs are of mesenchymal lineage and, when mature, produce osteoid, and the associated proteins, allowing for mineralization and subsequent maturation of bone. Bone lining cells are quiescent flat-shaped OBs that cover inactive bone surfaces. Bone lining cells can be induced to proliferate and differentiate into osteogenic cells and may represent a source of "determined" osteogenic precursors (Matic *et al.*, 2016; Miller *et al.*, 1989).

Osteocytes are end-differentiated OBs that are embedded and trapped in bone matrix. The formation of bone by OBs is coupled with OC degradation (Everts et al., 2002). OCs are derived from mature monocytes under a suitable micro-environment and are responsible for the breakdown of mineralized bone (Udagawa et al., 1990).

OC differentiation is regulated by OC differentiation factors, mainly macrophage colony stimulating factor (M-CSF) and receptor activator of nuclear factor NF-κB ligand (RANKL). M-CSF is produced by a number of cell types including immune cells (monocytes, activated macrophages, lymphocytes, and granulocytes), endothelial cells, fibroblasts, osteocytes, and OBs. Apoptotic osteocytes trigger signals, likely vascular endothelial growth factor (VEGF) (Cheung et al., 2011; Varoga et al., 2009) that attracts OC precursors (OCPs) to specific areas of bone, which in turn differentiate and fuse to become mature, bone-resorbing OCs (Bellido, 2014; Shandala et al., 2012; Tatsumi et al., 2007; Verborgt et al., 2000). Along with RANKL, interleukin-1 (IL-1), IL-6, IL-11, tumor necrosis factor-α (TNF-α), and TNF-β, or inhibitory factors like osteoprotegerin (Simonet et al., 1997; Tsuda et al., 1997), IL-3 (Khapli et al., 2010), IL-27 (Kalliolias et al., 2010), and IL-33 (Saleh et al., 2011; Schulze et al., 2011) are all involved in the regulation of OC differentiation. OBs, osteocytes, activated T-cells, neutrophils, fibroblasts, and chondrocytes (Usui et al., 2008) all produce RANKL. The expression of RANKL by these cells is initiated, promoted, or inhibited by osteotropic hormones and cytokines.

8.2. Rho GTPases and osteoclasts

Small GTPases behave as molecular switches and clocks controling signaling transduction in various cell types. The mammalian Rho family can be divided into eight subgroups comprising 20 members (Boureux et al., 2007). Generally speaking, Rho GTPases are regulators of the actin cytoskeleton, including the formation of filopodia, lamellipodia, membrane ruffles, and stress fibers (Razzouk et al., 1999; Ridley and Hall, 1992; Ridley et al., 1992). These structures therefore play regulatory roles in the formation and function of OCs.

During OC formation, OCPs undergo a number of steps to become multi-nucleated cells. M-CSF enables survival, proliferation, and recruitment of OCPs (Yavropoulou and Yovos, 2008). RANKL commits OCPs to mononucleated, tartrate-resistant acid phosphatase (TRAcP)-positive prefusion OCs (pre-OCs) (Kukita et al., 1993). These fusion-competent pre-OCs migrate and extend filopodia toward fusion partners to bring their membranes into close contact with one another (Helming and Gordon, 2009). Filopodia between fusion-competent cells allow for cell fusion and subsequent sharing of cytoplasmic contents.

8.3. Rho GTPases and osteoclast precursor migration

Our knowledge of the specific roles played by Rho GTPases in OC precursor cells is only beginning to be elucidated. As expected, those cellular processes that are affected by cytoskeletal organization, in particular actin organization, are the most likely to have Rho GTPases involvement.

Actin filaments are the core of podosomes, or small plasma-membrane protrusions, that are located on the ventral side of a number of cells, including monocyte-derived cells, such as macrophages (Davies and Stossel, 1977), OCs (Marchisio et al., 1984), and dendritic cells (Binks et al., 1998). The main functions of podosomes appear to be cell adhesion, motility, substrate recognition, and degradation of the extracellular matrix (Burgstaller and Gimona, 2005; Chellaiah et al., 2000a; Osiak et al., 2005; Teti et al., 1991). In vitro and in vivo experiments provide clear evidence that the most important role played by Rho GTPases is of regulation of podosome assembly and organization (Lakkakorpi and Vaananen, 1991).

The most prominently expressed Rho GTPases in human myeloid cells from the granulocyte–monocyte lineage are Cdc42, RhoQ, Rac1, Rac2, RhoA, RhoC, RhoF, and RhoV on a mRNA level (van Helden et al., 2012). RhoA, Rac1, and Cdc42 are critical in podosome rearrangement, cell migration, and OC function (Chellaiah et al., 2000b). It has been shown that inhibition of Rho by a C3-derived fusion protein in RAW264.7, a M-CSF expressing monocytic cell line, inhibits cell proliferation (Tautzenberger et al., 2013). Effective cell migration of OCPs occurs through spatiotemporally coordinated activation of RhoA, Rac1, and Cdc42 (Kim et al., 2016; Ory et al., 2008). Rac and M-CSF-activating Cdc42 GTPases control hematopoietic stem cell shape, as well as adhesion, migration, and mobilization (Ito et al., 2010; Yang et al., 2001). Deletion of either Rac1 or Rac2 results in no change in the total number of macrophage colony-forming unit when compared to wild-type mice. However, when both Rac1 and Rac2 are deleted, a significantly high amount of OCPs are seen (Wang et al., 2008).

Expression of RhoU/Wrch1 by RAW264.7 cells was shown to favor cell aggregation, indicating the importance of the control of OCP migratory properties, through modulation of $\alpha_v\beta_3$ integrin signaling to regulate OC precursor adhesion and migration (Brazier et al., 2009). Interestingly, RhoU/Wrch1 is a negative modulator of OCP migration in response to M-CSF, indicating different attraction molecules must act on OCPs at different stages.

The importance of Rho GTPases has also been elucidated in OCPs during migration, interestingly, with respect to the formation of reactive oxygen species (ROS). Nicotinamide adenine dinucleotide phosphate (NADPH) oxidase-generated ROS act as an intracellular signal mediator for RANKL-induced OC differentiation and fusion (Lee *et al.*, 2005) (for review, see Quinn and Schepetkin (2009)). Although ROS are generally known as side products of oxidative respiration that kill bacteria and cause cellular damage, there is an expanding body of literature demonstrating that ROS-mediated oxidation of proteins may be an important signal-transduction mechanism in many cellular processes including cell migration (Chiarugi, 2005). Superoxide is the primary ROS produced by cellular NADPH oxidases (NOX) and via spontaneous or enzymatic dismutation gives rise to the ROS hydrogen peroxide, a key signal mediator in many cell types (Finkel, 2000). Through oxidation, the ROS hydrogen peroxide can alter the biological characteristics of susceptible protein targets, affecting their enzymatic function and/or binding characteristics. Epidemiological evidence in humans and studies in rodents indicate that ROS are the primary culprits associated with increased generation of OCs in hormone and age-related osteoporosis (Manolagas, 2010). The main source of ROS in OCs is NOX2, which consists of two integral membrane subunits (gp91 and p22) and four subunits (p67, p40 p47, and Rac) that are recruited from the cytosol (Bokoch and Zhao, 2006). It is known that Rac association with NOX1 and NOX2 is required for oxidase assembly (Kim and Dinauer, 2001). There is evidence suggesting that NOX1 and NOX4 may also be up-regulated in mature OCs (Yang *et al.*, 2004), although the relative contribution and roles of NOX1, NOX2, and NOX4 in ROS production during OC fusion remains to be determined.

8.4. Rho GTPases and prefusion-osteoclast fusopodia

The actin cytoskeleton is also very important, not only for the motility of OCs at all differentiation stages, but also in the process of pre-OCs fusion to one another through fusopods (Wang *et al.*, 2015). Fusopods are filipodial extensions that are rich in actin. The plasma membrane in other fusing cell types has been shown to be subdivided into several distinct microdomains or rafts containing fusogenic proteins (Larsson, 2011). Therefore, the fusopodial cell membrane structure is proposed to serve as a platform for fusogen-rich proteins. Cells moving towards one another in preparation for fusion have been observed in real time to extend filipodia and actin rings at the leading edge (Hu *et al.*, 2011).

Rac1 plays an important role in this process along with its downstream ROS signaling. Rac1, in general, promotes lamellipodia and focal-complex formation,

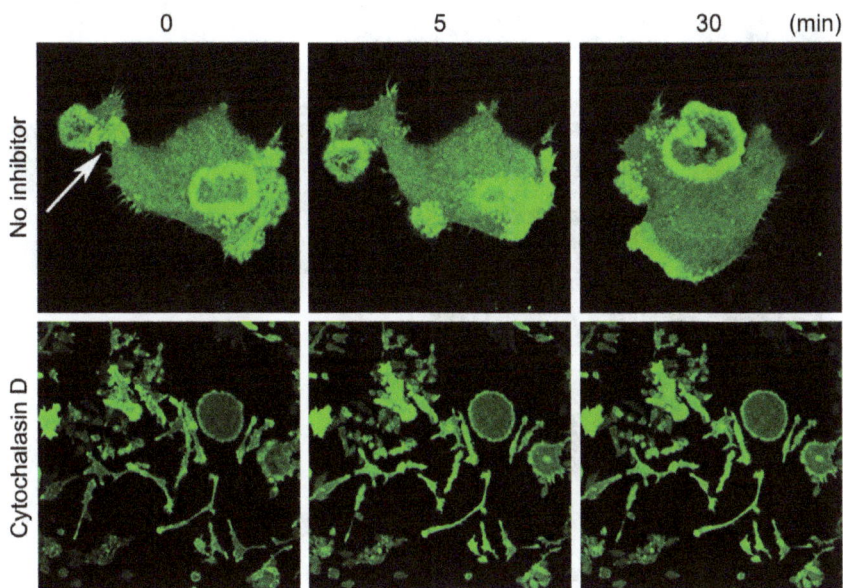

Figure 1. Time-lapse imaging shows cytochalasin D-inhibited cell fusion (original magnification 40×). pmEGFP-N1-Lifeact (provided by Dr. Roland Wedlich-Soldner, Cellular Dynamics and Cell Patterning, Max-Planck Institute of Biochemistry, Martinsried, Germany) (Riedl *et al.*, 2008) stable transfected RAW264.7 cells were stimulated with soluble RANKL (sRANKL, 60 ng/ml) for 2 days. The cells were RANKL starved overnight. The cells were either treated with dimethyl sulfoxide (DMSO) (control) or cytochalasin D (100 nM) for 30 min before sRANKL reloading and confocal live cell imaging. The microscope was equipped with a heating stage. CO_2 premix (5%) was provided. First row shows control cells from contact (arrow pointed) to fusion completed within 30 min, whereas no fusion events were observed in cells treated with cytochalasin D.

and Cdc42 induces actin polymerization to form filopodia (Ory *et al.*, 2008). Actin polarization inhibitor cytochalasin D blocks cell fusion (Fig. 1), indicating that the actin cytoskeleton is needed for not only cell migration but also fusopod formation.

Rac1-null pre-OCs generated fewer fusopods under RANKL stimulation (Wang *et al.*, 2015). This has been confirmed by the finding that cell extension formation was impaired by the arrest of Rac1 in response to M-CSF and hepatocyte growth factor (HGF) (Dumontier *et al.*, 2000; Faccio *et al.*, 2003). Additionally, the suppression or deletion of Rac1 in mouse monocytes resulted in reduced OC fusion by affecting downstream ROS signaling (Lee *et al.*, 2005; Wang *et al.*, 2008). The blockage of ROS production with N-acetyl-L-cysteine

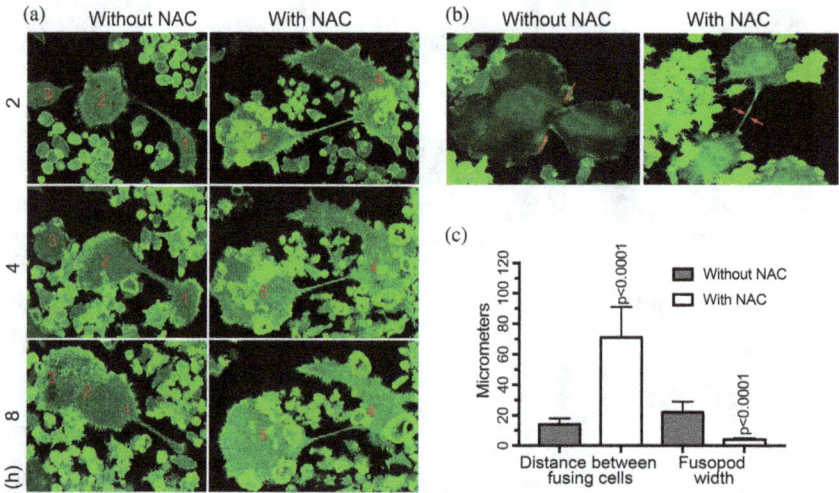

Figure 2. Cell fusion was blocked by NAC (original magnification 40×). (a) NAC (25 mM) was added to the cultures 30 min before sRANKL reloading. sRANKL was added into the cultures, and the cells were submitted to a confocal microscope for live cell imaging (10-hour period). The cells (Nos. 1, 2, and 3) without NAC treatment fused; however, cells (Nos. 4 and 5) treated with NAC did not fuse. Note that actin waves appeared in the control (no NAC) fusing cells and disappeared in the fusing cells treated with NAC. (b) The stable cells were treated with 60 ng/ml sRANKL in a six-well plate on a 25CIR cover glass for 2 days. The cells were RANKL starved overnight. NAC (25 mM) was added to the cultures 30 min before sRANKL restimulation. The cells were restimulated with sRANKL for 2 h and fixed. (c) Statistical analysis of data obtained from experiment B shows fusopod (red arrow pointed) length and width were inhibited by NAC.

(NAC) dramatically inhibits OC fusion events through disruption of normal actin dynamics (Fig. 2).

8.5. Rho GTPases in the early stage of osteoclast differentiation

The differentiation of OCPs, along with the multinucleation process previously mentioned, are also important to consider during osteoclastogenesis (OCG), as well as the role that Rho GTPases play. RANKL has been shown to activate Cdc42 as previously mentioned (Kim *et al.*, 2009a). Activated Cdc42 is involved in several osteoclastogenic signaling events in the earlier stage of OC differentiation, including expression and activation of the differentiation factors microphthalmia-associated transcription factor (MiTF) and nuclear factor of activated T-cell, cytoplasmic 1 (NFATc1) (Ito et al., 2010). RANKL-induced OCG shows up-regulation of RhoU/Wrch1 at both mRNA and protein levels, the only Rho family of small GTPases during OCG. However, it is not essential for the establishment of the entire OCG transcriptional program to produce the characteristic markers of osteoclastic differentiation such as NFATc1, Src, TRAcP, and cathepsin K (Brazier et al., 2009). Also as described previously, RANKL-induced ROS production is observed to increase the generation of OCs (Fig. 3) (Gambari et al., 2014; Ha et al., 2004; Kanzaki et al., 2013; Moon et al., 2012; Steinbeck et al., 1998; Suda et al., 1993; Zhou et al., 2013).

It is now known that cofilin, a key actin regulatory protein, is itself regulated by ROS-mediated liberation of phosphatase, slingshot (Kim et al., 2009b). Rac1 plays a role in OCG, in part due to ROS generation (Kim et al., 2010). RANKL stimulation of bone marrow monocyte–macrophage lineage cells transiently increases the intracellular level of ROS through a signaling cascade involving TRAF6, Rac1, and NADPH oxidase (Lee et al., 2005). The increased level of ROS was shown to act as an intracellular signal mediator of OCG, although the exact mechanisms are not yet fully characterized (Lee et al., 2005).

8.6. Rho GTPases and prefusion-osteoclast fusion

It is hypothesized that the plasma membrane of both fusing cells, in general, fuse via the tethering of aqueous pores and expansion of these pores, leading to the merging of the cytoplasm (Vignery, 2011). Similar to myoblast fusion (for review, see Chen et al. (2007)), OC fusion is initiated by the interaction of cells though membrane extensions (Song et al., 2014b) or fusopods (Wang et al., 2015) (also see Fig. 4). The formation of cell pseudopods, as well as direct sinking of the plasma membrane of one cell to another, has been shown in the IL-4-induced foreign body giant cell formation (Dugast et al., 1997; McNally and Anderson, 2005). Cell extensions contact the fusion partner and are followed by a series of actin waves, which stabilize and increase the girth of the fusopod prior to fusion. Studies looking at macrophage and syncytia fusion show that disruption of the actin cytoskeleton by cytochalasins and latrunculin blocked cell fusion events, clearly implicating an intact actin cytoskeleton in cell fusion (DeFife et al., 1999; Pine et al., 1998). It is not surprising that there is a requirement for Rac1 for OCG (Lee et al., 2006; Wang et al., 2015; Wang et al., 2008), as Rho GTPases are cytoskeleton regulators (Ridley and Hall, 1992; Ridley et al., 1992) and filamentous actin

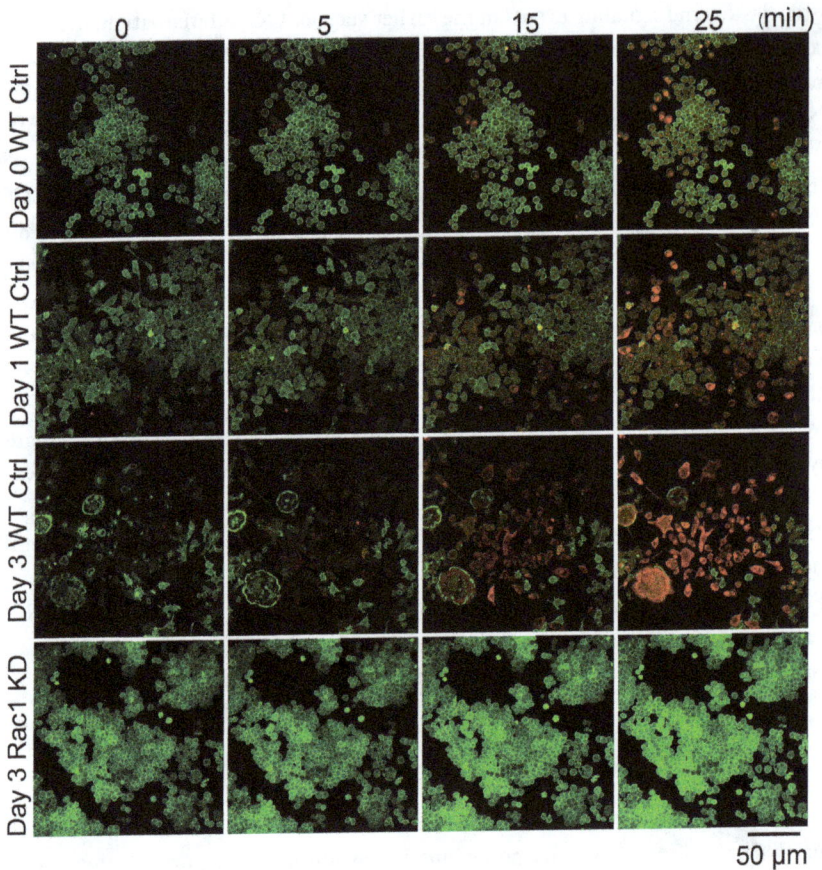

Figure 3. ROS production during osteoclast differentiation. Lifeact-mEGFP stable transfected OC precursor cell line RAW264.7 cells were stimulated with sRANKL (60 ng/mL) for 1–3 days. For cells that were stimulated for 3 days, the medium was replaced at Day 2 with fresh medium containing sRANKL. On the day of assays, dihydroethidium (DHE) (0.8 μM/mL) was loaded into cell cultures and submitted immediately to confocal microscope. Time-lapse imaging was taken every 30 seconds for a total period of 30 min. It took 24 ± 4, 5 ± 1, or 6 ± 2 min for RANKL-unstimulated wild-type control (WT Ctrl) cells, Day 1 or Day 3 RANKL-stimulated WT Ctrl cells to turn red (ethidium, ETH), respectively, after DHE addition. There was no ROS production observed by RANKL in Rac1 stable knockdown (KD) cells (Wang *et al.*, 2015) within this period of assessment.

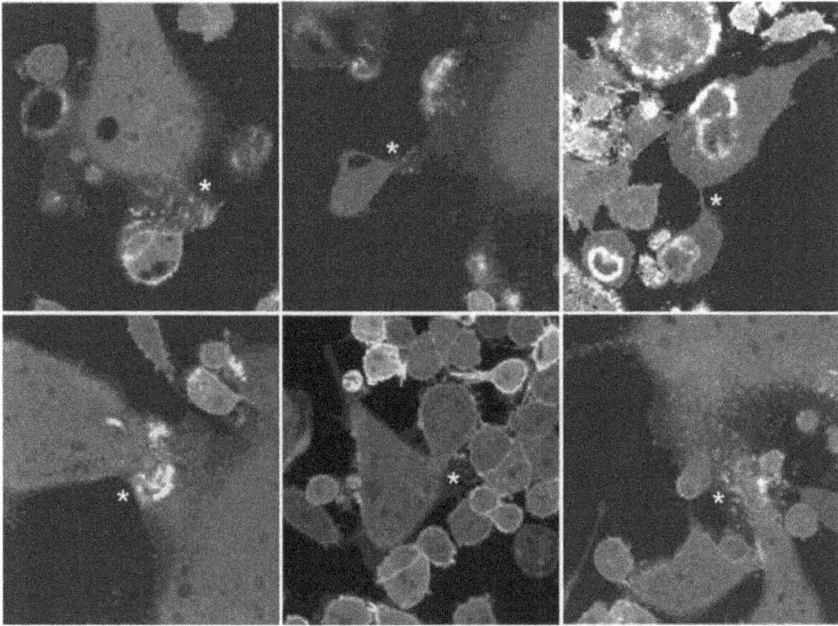

Figure 4. Cell fusion is mediated by fusopods (original magnification 40×). pmEGFP-Lifeact stably transfected RAW264.7 cells were stimulated with sRANKL for 2 days. The cells were RANKL starved overnight. sRANKL was reloaded, and cell fusion events were captured (Wang *et al.*, 2015). Pictures show cell fusion was mediated by cell membrane extensions (star pointed).

(F-actin)-enriched plasma membrane protrusions are involved in cell fusion (Kim *et al.*, 2007).

It is unknown through which mechanisms the small GTPases actually regulate cell fusion. Additionally, it has also been suggested that the actin cytoskeleton signaling network determines the size of OCs during cell fusion through Rho and Rac1 involvement (Takito *et al.*, 2015).

Blocking lysophosphatidic acid receptor type-1 (LPA-1) activity with Ki16425 inhibited expression of NFATc1 and dendritic cell-specific transmembrane protein and interfered with pre-OCs' fusion but not the proliferation of OCPs (David *et al.*, 2014). LPA-1 controls sealing zone formation acting though RhoA (David *et al.*, 2014). LPA-1 expression markedly increased in the bone of mice that had underwent bilateral oophorectomies, which was blocked by bisphosphonate treatment (David *et al.*, 2014). MiR-31 controls cytoskeleton organization in OCs for optimal bone resorption activity by regulating the expression of RhoA (Mizoguchi *et al.*, 2013). MicroRNA-124 (miR-124) regulates OCG of mouse bone marrow macrophages by the expression of RhoA and Rac1 in OCPs (Lee *et al.*, 2013).

8.7. Rho GTPase association with osteoclast polarization, vesicular transport, and migration

Active OCs are highly polarized cells. Their plasma membrane is organized into three distinct domains: a ruffled border, a sealing zone, and a functional secretory domain (FSD). The ruffled border is where protons and proteinases are secreted to degrade the bone matrix, which is endocytosed by the ruffled border, transcytosed via microtubules and released through an FSD. The activation of OCs is reflected in the podosome assembly and disassembly. The formation and disassembly of podosomes in OCs aids to group podosomes into clusters and rings in the immature OCs and form belts in the inactive mature OCs. OCs on the bone surface cycle between resorptive and migratory phases, with podosome rings generating forces that drive OC migration (Hu *et al.*, 2011). It was found that OC resorptive activity was suppressed by disruption of the actin skeleton in the sealing zone, a broad prominent band between membrane and the mineralized matrix in which radial actin fibers from the bounded F-actin core of single podosomes are connected (Destaing *et al.*, 2003). RhoA and Rac1 are known to be necessary to maintain F-actin organization in multinucleated giant cells (Chellaiah *et al.*, 2000b; Ory *et al.*, 2000). RhoA and Rac are essential for podosome assembly and resorbing function (Chellaiah, 2005; Razzouk *et al.*, 1999; Zhang *et al.*, 1995). Rho has also been described to be involved in OC actin cytoskeleton remodeling and vesicular transport to the FSD (Itzstein *et al.*, 2011). Rho GTPases do not induce podosome formation (for review, see Ory *et al.* (2008)) but aid in controlling podosome organization and reorganization. OC bone resorbing capacity is reduced by antioxidants that block ROS production downstream of Rac1 (Ha *et al.*, 2004). Rac1 is an effector of neurofibromin-1 (NF1), a tumor suppressor gene that functions as a GTPase-activating protein, in OCs (Yan *et al.*, 2008). *Nf1* heterozygous OCs demonstrated significantly higher levels of belt formation as compared to all other genotypes, while Rac1-null OCs demonstrated significantly less numbers of both belt and ring structures. Upon genetic deletion of *Rac1*, the *Nf1* heterozygous OCs demonstrated belt formation at a similar level to that of wild-type OCs (Yan *et al.*, 2008). Macrophage polykaryons have disorganized podosomes when activated or dominant-negative Rac1 are microinjected (Ory *et al.*, 2000). Expression of activated Cdc42 leads to podosome and podosome belt disruption in macrophages and OCs, respectively (Chellaiah, 2005; Linder *et al.*, 1999). Cdc42 is a component of the Par3/Par6/atypical PKC OC polarization complex, which governs the rate of actin ring formation (Ito *et al.*, 2010).

Rho activity is required for the formation of the apatite-dependent sealing zone and polarization, but RhoA activation in OCs cultured on glass coverslips produce a sealing zone (Destaing et al., 2005; Ory et al., 2000, 2008). The expression of RhoA is controlled by miR-31 (Mizoguchi et al., 2013) to avoid overexpression of RhoA or increased basal RhoA activity (Gil-Henn et al., 2007). Activation of RhoA in the resorbing OC on the mineral substrates has to reach an optimized level. RhoE is a novel regulator of actin dynamics in bone-resorbing OCs (Georgess et al., 2014). RhoU/Wrch1, however, is not required for sealing zone formation or bone resorption (Brazier et al., 2009) but is important for the localization of RhoU/Wrch1 in OC podosomes (Ory et al., 2007), indicating its involvement in OC migration, adhesion, and fusion. Rac1 plays a significant role in bone resorptive activity by regulating the motility of OCs as well (Fukuda et al., 2005). Lastly, it is thought that calcium concentration change in the resorption lacunae during resorption might be a pivotal regulator of Rho GTPase in OC polarization and depolarization.

8.8. Rho GTPases and osteoclast survival

Rho GTPases are crucial not only in OC differentiation and function but also in OC survival. Cdc42 is critical for OC survival and lifespan in vivo and in vitro (Ito et al., 2010). Rac1 (Fukuda et al., 2005), RhoA, and RhoV are all associated with antiapoptosis signaling. RhoV has been shown to mediate apoptosis of RAW264.7 macrophages during OC differentiation (Song et al., 2015). Prolonged Cdc42 activation and attenuated OCG in LIS1-null OCPs was due to promoted cell death (Ito et al., 2010; Ye et al., 2016). The absence of Cdc42 in mature OCs led to more rapid cell death than wild-type OCs upon withdrawal of M-CSF and RANKL (Ito et al., 2010). Antioxidants, such as NAC and glutathione, prevented RANKL-induced activation of Akt, NF-κB, and ERK, which are important factors in cytokine-stimulated OC survival (Ha et al., 2004). Additionally, Rac1 mediates survival signaling of OCs primarily by modulating PI3K/Akt pathways (Fukuda et al., 2005).

8.9. Factors regulating Rho GTPase activity in osteoclasts

Guanine nucleotide exchange factors (GEFs), GTPases-activating proteins (GAPs), GDP dissociation inhibitors (GDIs), GDI-like proteins (for review, see Cherfils and Zeghouf (2013) and Nayak et al. (2013)) are the regulators of Rho GTPase activity. Rho guanine nucleotide exchange factors (GEFs) Arhgef8/Net1 and Dock5 were upregulated upon RANKL stimulation (Brazier et al., 2006).

Knockdown Net1 or Dock5 mRNA expression in OCPs dramatically impaired OC formation, fusion, or function (Brazier *et al.*, 2006). Dock5 is an atypical Rac1 exchange factor. A study shows that $Dock5^{-/-}$ mice have increased trabecular bone mass with normal OC numbers, demonstrating Dock5 to be essential for sealing zone assembly and bone resorption but not OC differentiation (Vives *et al.*, 2011). The small chemical compound *N*-(3,5-dichlorophenyl) benzenesulfonamide (C21) directly inhibits the exchange activity of Dock5 and disrupts OC podosome organization. Remarkably, C21 administration protects mice against bone degradation in models recapitulating major osteolytic diseases such as rheumatoid arthritis and bone metastasis (Vives *et al.*, 2015). RhoGEFs as therapeutic targets is discussed in Chapter 9.

8.10. Rho GTPases as drug targets and future perspectives

A number of life-altering or life-threatening conditions are associated with bone loss, such as osteoporosis, rheumatoid arthritis, Paget's disease, periodontal disease, as well as metastatic and primary bone malignancies (D'Amico and Roato, 2012). Currently, OC activity and differentiation is inhibited through the use of bisphos-phonates, hormone replacement therapy, selective estrogen receptor modulators, calcitonin or biologic antibodies targeting RANKL, or vascular endothelial growth factor (Ji *et al.*, 2016; McClung *et al.*, 2006; Miller *et al.*, 2008). Nitrogen-containing bisphosphonates reduce OC activity by inhibiting the enzyme farnesyl pyrophos-phate synthetase (FPPS) and impairing the posttranslational modification of most of the Rho GTPases (Ras, Rab, and Rho families). Unfortunately, while inhibiting OCs and preventing bone loss, these medications have a number of unwanted side effects, and newer methodologies for OC inhibition are greatly sought after.

It was found that suppression of osteoclastogenic transcription factors c-Fos and NFATc1 expression inhibits RANKL-induced OC differentiation (Ha *et al.*, 2013a, b). Rho GTPases are critically involved in OC differentiation and function; therefore, the development of drugs that can selectively affect OCPs, OCG, and OC function, by targeting small GTPases and their related signaling pathways (Table 1), holds great promise.

The roles played by Rho GTPases in OC differentiation and functionality at all stages are incredibly important but varied. Involvement of Rho GTPases in actin cytoskeleton reorganization, including podosomes and fusopods, has drawn interest in the development of new drugs targeting Rho GTPases. Interruption of OC formation and activity would be beneficial in many pathological conditions, and therefore their precise mechanism of action in OCs and OCG is of great importance.

Table 1. Selected effector proteins for Rho GTPases in OC.

Member	Effector	Effect	Key Ref.
RhoA	mDia2	Podosome belt integrity; microtubule acetylation	Destaing et al. (2005)
	ROCK	Podosome compaction and sealing zone formation	Song et al. (2014a)
	PAK4	Acts to correctly localize PAK proteins to podosomes.	Abo et al. (1998); Gringel et al. (2006)
RhoU/Wrch1	$\alpha_v \beta_3$ integrin	Modification of cell–cell and cell–substratum adhesion properties; migration; podosome organization (increased RhoU/Wrch1 activity led to impaired podosome belt formation)	Brazier et al. (2009)
RhoU/Wrch1; Cdc 42; Rac1; Rac2	PAK1/2/3	Podosome and sealing zone formation	Razzouk et al. (1999)
RhoA; Rac1	PIP(5)k	Actin ring formation; bone resorption	Chellaiah (2005); Chong et al. (1994); Weernink et al. (2004); Zhu et al. (2013)
Rac1	P67[phox]	Osteoclast differentiation, fusion and function	Sasaki et al. (2009)
	WAVE complex	Sealing zone formation	Touaitahuata et al. (2014)
Rac1; Cdc42	FlnA	Osteoclast precursor cell migration	Leung et al. (2010)
Cdc42	WASp	Podosome formation and assembly; podosome disassembly	Calle et al. (2004); Chabadel et al. (2007); Chellaiah et al. (2000b); Chou et al. (2006)

8.11. Abbreviations

F-actin:	filamentous actin
FSD:	functional secretory domain
GAPs:	GTPase-activating proteins
GDIs:	GDP-dissociation inhibitors
GEFs:	guanine nucleotide exchange factors
IL:	interleukin
LPA:	lysophosphatidic acid receptor
M-CSF:	macrophage colony-stimulating factor
NADPH:	nicotinamide adenine dinucleotide phosphate
NFATc1:	nuclear factor of activated T-cell, cytoplasmic 1
NOX:	NADPH oxidase
OB:	osteoblast
OCP:	osteoclast precursor
OC:	osteoclast
OCG:	osteoclastogenesis
RANKL:	receptor activator of nuclear factor NF-κB ligand
ROS:	reactive oxygen species
TNF:	tumor necrosis factor

References

Abo, A., Qu, J., Cammarano, M.S., Dan, C., Fritsch, A., Baud, V., Belisle, B., and Minden, A. (1998). PAK4, a novel effector for Cdc42Hs, is implicated in the reorganization of the actin cytoskeleton and in the formation of filopodia. EMBO J. *17*, 6527–6540.

Bellido, T. (2014). Osteocyte-driven bone remodeling. Calcif. Tissue Int. *94*, 25–34.

Binks, M., Jones, G.E., Brickell, P.M., Kinnon, C., Katz, D.R., and Thrasher, A.J. (1998). Intrinsic dendritic cell abnormalities in Wiskott-Aldrich syndrome. Eur. J. Immunol. *28*, 3259–3267.

Bokoch, G.M. and Zhao, T. (2006). Regulation of the phagocyte NADPH oxidase by Rac GTPase. Antioxid. Redox Signal. *8*, 1533–1548.

Boureux, A., Vignal, E., Faure, S., and Fort, P. (2007). Evolution of the Rho family of ras-like GTPases in eukaryotes. Mol. Biol. Evol. *24*, 203–216.

Brazier, H., Pawlak, G., Vives, V., and Blangy, A. (2009). The Rho GTPase Wrch1 regulates osteoclast precursor adhesion and migration. Int. J. Biochem. Cell Biol. *41*, 1391–1401.

Brazier, H., Stephens, S., Ory, S., Fort, P., Morrison, N., and Blangy, A. (2006). Expression profile of RhoGTPases and RhoGEFs during RANKL-stimulated osteoclastogenesis: identification of essential genes in osteoclasts. J. Bone Miner. Res. *21*, 1387–1398.

Burgstaller, G. and Gimona, M. (2005). Podosome-mediated matrix resorption and cell motility in vascular smooth muscle cells. Am. J. Physiol. *288*, H3001–H3005.

Calle, Y., Jones, G.E., Jagger, C., Fuller, K., Blundell, M.P., Chow, J., Chambers, T., and Thrasher, A.J. (2004). WASp deficiency in mice results in failure to form osteoclast sealing zones and defects in bone resorption. Blood *103*, 3552–3561.

Chabadel, A., Banon-Rodriguez, I., Cluet, D., Rudkin, B.B., Wehrle-Haller, B., Genot, E., Jurdic, P., Anton, I.M., and Saltel, F. (2007). CD44 and beta3 integrin organize two functionally distinct actin-based domains in osteoclasts. Mol. Biol. Cell *18*, 4899–4910.

Chellaiah, M., Kizer, N., Silva, M., Alvarez, U., Kwiatkowski, D., and Hruska, K.A. (2000a). Gelsolin deficiency blocks podosome assembly and produces increased bone mass and strength. J. Cell Biol. *148*, 665–678.

Chellaiah, M.A. (2005). Regulation of actin ring formation by rho GTPases in osteoclasts. J. Biol. Chem. *280*, 32930–32943.

Chellaiah, M.A., Soga, N., Swanson, S., McAllister, S., Alvarez, U., Wang, D., Dowdy, S.F., and Hruska, K.A. (2000b). Rho-A is critical for osteoclast podosome organization, motility, and bone resorption. J. Biol. Chem. *275*, 11993–12002.

Chen, E.H., Grote, E., Mohler, W., and Vignery, A. (2007). Cell-cell fusion. FEBS Lett. *581*, 2181–2193.

Cherfils, J. and Zeghouf, M. (2013). Regulation of small GTPases by GEFs, GAPs, and GDIs. Physiol. Rev. *93*, 269–309.

Cheung, W.Y., Liu, C., Tonelli-Zasarsky, R.M., Simmons, C.A., and You, L. (2011). Osteocyte apoptosis is mechanically regulated and induces angiogenesis *in vitro*. J. Orthop. Res. *29*, 523–530.

Chiarugi, P. (2005). PTPs versus PTKs: the redox side of the coin. Free Radic. Res. *39*, 353–364.

Chong, L.D., Traynor-Kaplan, A., Bokoch, G.M., and Schwartz, M.A. (1994). The small GTP-binding protein Rho regulates a phosphatidylinositol 4-phosphate 5-kinase in mammalian cells. Cell *79*, 507–513.

Chou, H.C., Anton, I.M., Holt, M.R., Curcio, C., Lanzardo, S., Worth, A., Burns, S., Thrasher, A.J., Jones, G.E., and Calle, Y. (2006). WIP regulates the stability and localization of WASP to podosomes in migrating dendritic cells. Curr. Biol. *16*, 2337–2344.

D'Amico, L. and Roato, I. (2012). Cross-talk between T cells and osteoclasts in bone resorption. Bonekey Rep. *1*, 82.

David, M., Machuca-Gayet, I., Kikuta, J., Ottewell, P., Mima, F., Leblanc, R., Bonnelye, E., Ribeiro, J., Holen, I., Lopez Vales, R., *et al.* (2014). Lysophosphatidic acid receptor type 1 (LPA1) plays a functional role in osteoclast differentiation and bone resorption activity. J. Biol. Chem. *289*, 6551–6564.

Davies, W.A. and Stossel, T.P. (1977). Peripheral hyaline blebs (podosomes) of macrophages. J. Cell Biol. *75*, 941–955.

DeFife, K.M., Jenney, C.R., Colton, E., and Anderson, J.M. (1999). Disruption of filamentous actin inhibits human macrophage fusion. FASEB J. *13*, 823–832.

Destaing, O., Saltel, F., Geminard, J.C., Jurdic, P., and Bard, F. (2003). Podosomes display actin turnover and dynamic self-organization in osteoclasts expressing actin-green fluorescent protein. Mol. Biol. Cell *14*, 407–416.

Destaing, O., Saltel, F., Gilquin, B., Chabadel, A., Khochbin, S., Ory, S., and Jurdic, P. (2005). A novel Rho-mDia2-HDAC6 pathway controls podosome patterning through microtubule acetylation in osteoclasts. J. Cell Sci. *118*, 2901–2911.

Dugast, C., Gaudin, A., and Toujas, L. (1997). Generation of multinucleated giant cells by culture of monocyte-derived macrophages with IL-4. J. Leukoc. Biol. *61*, 517–521.

Dumontier, M., Hocht, P., Mintert, U., and Faix, J. (2000). Rac1 GTPases control filopodia formation, cell motility, endocytosis, cytokinesis and development in Dictyostelium. J. Cell Sci. *113 (Pt 12)*, 2253–2265.

Everts, V., Delaisse, J.M., Korper, W., Jansen, D.C., Tigchelaar-Gutter, W., Saftig, P., and Beertsen, W. (2002). The bone lining cell: its role in cleaning Howship's lacunae and initiating bone formation. J. Bone Miner. Res. *17*, 77–90.

Faccio, R., Novack, D.V., Zallone, A., Ross, F.P., and Teitelbaum, S.L. (2003). Dynamic changes in the osteoclast cytoskeleton in response to growth factors and cell attachment are controlled by beta3 integrin. J. Cell Biol. *162*, 499–509.

Finkel, T. (2000). Redox-dependent signal transduction. FEBS Lett. *476*, 52–54.

Fukuda, A., Hikita, A., Wakeyama, H., Akiyama, T., Oda, H., Nakamura, K., and Tanaka, S. (2005). Regulation of osteoclast apoptosis and motility by small GTPase binding protein Rac1. J. Bone Miner. Res. *20*, 2245–2253.

Gambari, L., Lisignoli, G., Cattini, L., Manferdini, C., Facchini, A., and Grassi, F. (2014). Sodium hydrosulfide inhibits the differentiation of osteoclast progenitor cells via NRF2-dependent mechanism. Pharmacol. Res. *87*, 99–112.

Georgess, D., Mazzorana, M., Terrado, J., Delprat, C., Chamot, C., Guasch, R.M., Perez-Roger, I., Jurdic, P., and Machuca-Gayet, I. (2014). Comparative transcriptomics reveals RhoE as a novel regulator of actin dynamics in bone-resorbing osteoclasts. Mol. Biol. Cell *25*, 380–396.

Gil-Henn, H., Destaing, O., Sims, N.A., Aoki, K., Alles, N., Neff, L., Sanjay, A., Bruzzaniti, A., De Camilli, P., Baron, R., *et al.* (2007). Defective microtubule-dependent podosome organization in osteoclasts leads to increased bone density in Pyk2(−/−) mice. J. Cell Biol. *178*, 1053–1064.

Gringel, A., Walz, D., Rosenberger, G., Minden, A., Kutsche, K., Kopp, P., and Linder, S. (2006). PAK4 and alphaPIX determine podosome size and number in macrophages through localized actin regulation. J. Cell. Physiol. *209*, 568–579.

Ha, H., An, H., Shim, K.S., Kim, T., Lee, K.J., Hwang, Y.H., and Ma, J.Y. (2013a). Ethanol extract of *Atractylodes macrocephala* protects bone loss by inhibiting osteoclast differentiation. Molecules *18*, 7376–7388.

Ha, H., Kwak, H.B., Lee, S.W., Jin, H.M., Kim, H.M., Kim, H.H., and Lee, Z.H. (2004). Reactive oxygen species mediate RANK signaling in osteoclasts. Exp. Cell Res. *301*, 119–127.

Ha, H., Shim, K.S., An, H., Kim, T., and Ma, J.Y. (2013b). Water extract of *Spatholobus suberectus* inhibits osteoclast differentiation and bone resorption. BMC Complement. Altern. Med. *13*, 112.

Helming, L. and Gordon, S. (2009). Molecular mediators of macrophage fusion. Trends Cell Biol. *19*, 514–522.

Hu, S., Planus, E., Georgess, D., Place, C., Wang, X., Albiges-Rizo, C., Jurdic, P., and Geminard, J.C. (2011). Podosome rings generate forces that drive saltatory osteoclast migration. Mol. Biol. Cell *22*, 3120–3126.

Ito, Y., Teitelbaum, S.L., Zou, W., Zheng, Y., Johnson, J.F., Chappel, J., Ross, F.P., and Zhao, H. (2010). Cdc42 regulates bone modeling and remodeling in mice by modulating RANKL/M-CSF signaling and osteoclast polarization. J. Clin. Invest. *120*, 1981–1993.

Itzstein, C., Coxon, F.P., and Rogers, M.J. (2011). The regulation of osteoclast function and bone resorption by small GTPases. Small GTPases *2*, 117–130.

Ji, X., Chen, X., and Yu, X. (2016). MicroRNAs in osteoclastogenesis and function: potential therapeutic targets for osteoporosis. Int. J. Mol. Sci. *17*, 349.

Kalliolias, G.D., Zhao, B., Triantafyllopoulou, A., Park-Min, K.H., and Ivashkiv, L.B. (2010). Interleukin-27 inhibits human osteoclastogenesis by abrogating RANKL-mediated induction of nuclear factor of activated T cells c1 and suppressing proximal RANK signaling. Arthritis Rheumatol. *62*, 402–413.

Kanzaki, H., Shinohara, F., Kajiya, M., and Kodama, T. (2013). The Keap1/Nrf2 protein axis plays a role in osteoclast differentiation by regulating intracellular reactive oxygen species signaling. J. Biol. Chem. *288*, 23009–23020.

Khapli, S.M., Tomar, G.B., Barhanpurkar, A.P., Gupta, N., Yogesha, S.D., Pote, S.T., and Wani, M.R. (2010). Irreversible inhibition of RANK expression as a possible mechanism for IL-3 inhibition of RANKL-induced osteoclastogenesis. Biochem. Biophys. Res. Commun. *399*, 688–693.

Kim, C. and Dinauer, M.C. (2001). Rac2 is an essential regulator of neutrophil nicotinamide adenine dinucleotide phosphate oxidase activation in response to specific signaling pathways. J. Immunol. *166*, 1223–1232.

Kim, H., Choi, H.K., Shin, J.H., Kim, K.H., Huh, J.Y., Lee, S.A., Ko, C.Y., Kim, H.S., Shin, H.I., Lee, H.J., *et al.* (2009a). Selective inhibition of RANK blocks osteoclast maturation and function and prevents bone loss in mice. J. Clin. Invest. *119*, 813–825.

Kim, J.M., Kim, M.Y., Lee, K., and Jeong, D. (2016). Distinctive and selective route of PI3K/PKCalpha-PKCdelta/RhoA-Rac1 signaling in osteoclastic cell migration. Mol. Cell Endocrinol. *437*, 261–267.

Kim, J.S., Huang, T.Y., and Bokoch, G.M. (2009b). Reactive oxygen species regulate a slingshot-cofilin activation pathway. Mol. Biol. Cell *20*, 2650–2660.

Kim, M.S., Yang, Y.M., Son, A., Tian, Y.S., Lee, S.I., Kang, S.W., Muallem, S., and Shin, D.M. (2010). RANKL-mediated reactive oxygen species pathway that induces long lasting Ca2+ oscillations essential for osteoclastogenesis. J. Biol. Chem. *285*, 6913–6921.

Kim, S., Shilagardi, K., Zhang, S., Hong, S.N., Sens, K.L., Bo, J., Gonzalez, G.A., and Chen, E.H. (2007). A critical function for the actin cytoskeleton in targeted exocytosis of prefusion vesicles during myoblast fusion. Dev. Cell *12*, 571–586.

Kukita, A., Kukita, T., Shin, J.H., and Kohashi, O. (1993). Induction of mononuclear precursor cells with osteoclastic phenotypes in a rat bone marrow culture system depleted of stromal cells. Biochem. Biophys. Res. Commun. *196*, 1383–1389.

Lakkakorpi, P.T. and Vaananen, H.K. (1991). Kinetics of the osteoclast cytoskeleton during the resorption cycle in vitro. J. Bone Miner. Res. *6*, 817–826.

Larsson, L.-I. (2011). Regulation and control of cell–cell fusions. In: Larsson, L.-I., ed., *Cell Fusions: Regulation and Control*, Springer, New York, pp. 1–9.

Lee, N.K., Choi, H.K., Kim, D.K., and Lee, S.Y. (2006). Rac1 GTPase regulates osteoclast differentiation through TRANCE-induced NF-kappa B activation. Mol. Cell Biochem. *281*, 55–61.

Lee, N.K., Choi, Y.G., Baik, J.Y., Han, S.Y., Jeong, D.W., Bae, Y.S., Kim, N., and Lee, S.Y. (2005). A crucial role for reactive oxygen species in RANKL-induced osteoclast differentiation. Blood *106*, 852–859.

Lee, Y., Kim, H.J., Park, C.K., Kim, Y.G., Lee, H.J., Kim, J.Y., and Kim, H.H. (2013). MicroRNA-124 regulates osteoclast differentiation. Bone *56*, 383–389.

Leung, R., Wang, Y., Cuddy, K., Sun, C., Magalhaes, J., Grynpas, M., and Glogauer, M. (2010). Filamin A regulates monocyte migration through Rho small GTPases during osteoclastogenesis. J. Bone Miner. Res. *25*, 1077–1091.

Linder, S., Nelson, D., Weiss, M., and Aepfelbacher, M. (1999). Wiskott-Aldrich syndrome protein regulates podosomes in primary human macrophages. Proc. Natl. Acad. Sci. USA *96*, 9648–9653.

Manolagas, S.C. (2010). From estrogen-centric to aging and oxidative stress: a revised perspective of the pathogenesis of osteoporosis. Endocr. Rev. *31*, 266–300.

Marchisio, P.C., Cirillo, D., Naldini, L., Primavera, M.V., Teti, A., and Zambonin-Zallone, A. (1984). Cell-substratum interaction of cultured avian osteoclasts is mediated by specific adhesion structures. J. Cell Biol. *99*, 1696–1705.

Matic, I., Matthews, B.G., Wang, X., Dyment, N.A., Worthley, D.L., Rowe, D.W., Grcevic, D., and Kalajzic, I. (2016). Quiescent bone lining cells are a major source of osteoblasts during adulthood. Stem Cells *34*, 2930–2942.

McClung, M.R., Lewiecki, E.M., Cohen, S.B., Bolognese, M.A., Woodson, G.C., Moffett, A.H., Peacock, M., Miller, P.D., Lederman, S.N., Chesnut, C.H., *et al.* (2006). Denosumab in postmenopausal women with low bone mineral density. New Engl. J. Med. *354*, 821–831.

McNally, A.K. and Anderson, J.M. (2005). Multinucleated giant cell formation exhibits features of phagocytosis with participation of the endoplasmic reticulum. Exp. Mol. Pathol. *79*, 126–135.

Miller, P.D., Bolognese, M.A., Lewiecki, E.M., McClung, M.R., Ding, B., Austin, M., Liu, Y., San Martin, J., and Amg Bone Loss Study, G. (2008). Effect of denosumab on bone density and turnover in postmenopausal women with low bone mass after long-term continued, discontinued, and restarting of therapy: a randomized blinded phase 2 clinical trial. Bone *43*, 222–229.

Miller, S.C., de Saint-Georges, L., Bowman, B.M., and Jee, W.S. (1989). Bone lining cells: structure and function. Scanning Microsc. *3*, 953–960 (discussion 960-1).

Mizoguchi, F., Murakami, Y., Saito, T., Miyasaka, N., and Kohsaka, H. (2013). mir-31 controls osteoclast formation and bone resorption by targeting RhoA. Arthritis Res. Ther. *15*, R102.

Moon, H.J., Ko, W.K., Han, S.W., Kim, D.S., Hwang, Y.S., Park, H.K., and Kwon, I.K. (2012). Antioxidants, like coenzyme Q10, selenite, and curcumin, inhibited osteoclast differentiation by suppressing reactive oxygen species generation. Biochem. Biophys. Res. Commun. *418*, 247–253.

Nayak, R.C., Chang, K.H., Vaitinadin, N.S., and Cancelas, J.A. (2013). Rho GTPases control specific cytoskeleton-dependent functions of hematopoietic stem cells. Immunol. Rev. *256*, 255–268.

Ory, S., Brazier, H., and Blangy, A. (2007). Identification of a bipartite focal adhesion localization signal in RhoU/Wrch-1, a Rho family GTPase that regulates cell adhesion and migration. Biol. Cell *99*, 701–716.

Ory, S., Brazier, H., Pawlak, G., and Blangy, A. (2008). Rho GTPases in osteoclasts: orchestrators of podosome arrangement. Eur. J. Cell Biol. *87*, 469–477.

Ory, S., Munari-Silem, Y., Fort, P., and Jurdic, P. (2000). Rho and Rac exert antagonistic functions on spreading of macrophage-derived multinucleated cells and are not required for actin fiber formation. J. Cell Sci. *113 (Pt 7)*, 1177–1188.

Osiak, A.E., Zenner, G., and Linder, S. (2005). Subconfluent endothelial cells form podosomes downstream of cytokine and RhoGTPase signaling. Exp. Cell Res. *307*, 342–353.

Pine, P.S., Weaver, J.L., Oravecz, T., Pall, M., Ussery, M., and Aszalos, A. (1998). A semi-automated fluorescence-based cell-to-cell fusion assay for gp120-gp41 and CD4 expressing cells. Exp. Cell Res. *240*, 49–57.

Quinn, M.T. and Schepetkin, I.A. (2009). Role of NADPH oxidase in formation and function of multinucleated giant cells. J. Innate Immun. *1*, 509–526.

Razzouk, S., Lieberherr, M., and Cournot, G. (1999). Rac-GTPase, osteoclast cytoskeleton and bone resorption. Eur. J. Cell Biol. *78*, 249–255.

Ridley, A.J., and Hall, A. (1992). The small GTP-binding protein rho regulates the assembly of focal adhesions and actin stress fibers in response to growth factors. Cell *70*, 389–399.

Ridley, A.J. Paterson, H.F., Johnston, C.L., Diekmann, D., and Hall, A. (1992). The small GTP-binding protein rac regulates growth factor-induced membrane ruffling. Cell *70*, 401–410.

Riedl, J., Crevenna, A.H., Kessenbrock, K., Yu, J.H., Neukirchen, D., Bista, M., Bradke, F., Jenne, D., Holak, T.A., Werb, Z., et al. (2008). Lifeact: a versatile marker to visualize F-actin. Nat. Methods *5*, 605–607.

Saleh, H., Eeles, D., Hodge, J.M., Nicholson, G.C., Gu, R., Pompolo, S., Gillespie, M.T., and Quinn, J.M. (2011). Interleukin-33, a target of parathyroid hormone and oncostatin m, increases osteoblastic matrix mineral deposition and inhibits osteoclast formation in vitro. Endocrinology *152*, 1911–1922.

Sasaki, H., Yamamoto, H., Tominaga, K., Masuda, K., Kawai, T., Teshima-Kondo, S., and Rokutan, K. (2009). NADPH oxidase-derived reactive oxygen species are essential for differentiation of a mouse macrophage cell line (RAW264.7) into osteoclasts. J. Med. Invest. *56*, 33–41.

Schulze, J., Bickert, T., Beil, F.T., Zaiss, M.M., Albers, J., Wintges, K., Streichert, T., Klaetschke, K., Keller, J., Hissnauer, T.N., et al. (2011). Interleukin-33 is expressed in differentiated osteoblasts and blocks osteoclast formation from bone marrow precursor cells. J. Bone Miner. Res. *26*, 704–717.

Shandala, T., Shen Ng, Y., Hopwood, B., Yip, Y.C., Foster, B.K., and Xian, C.J. (2012). The role of osteocyte apoptosis in cancer chemotherapy-induced bone loss. J. Cell Physiol. *227*, 2889–2897.

Simonet, W.S., Lacey, D.L., Dunstan, C.R., Kelley, M., Chang, M.S., Luthy, R., Nguyen, H.Q., Wooden, S., Bennett, L., Boone, T., et al. (1997). Osteoprotegerin: a novel secreted protein involved in the regulation of bone density. Cell *89*, 309–319.

Song, R., Gu, J., Liu, X., Zhu, J., Wang, Q., Gao, Q., Zhang, J., Cheng, L., Tong, X., Qi, X., et al. (2014a). Inhibition of osteoclast bone resorption activity through osteoprotegerin-induced damage of the sealing zone. Int. J. Mol. Med. *34*, 856–862.

Song, R., Liu, X., Zhu, J., Gao, Q., Wang, Q., Zhang, J., Wang, D., Cheng, L., Hu, D., Yuan, Y., et al. (2015). RhoV mediates apoptosis of RAW264.7 macrophages caused by osteoclast differentiation. Mol. Med. Rep. *11*, 1153–1159.

Song, R.L., Liu, X.Z., Zhu, J.Q., Zhang, J.M., Gao, Q., Zhao, H.Y., Sheng, A.Z., Yuan, Y., Gu, J.H., Zou, H., et al. (2014b). New roles of filopodia and podosomes in the differentiation and fusion process of osteoclasts. Genet. Mol. Res. *13*, 4776–4787.

Steinbeck, M.J., Kim, J.K., Trudeau, M.J., Hauschka, P.V., and Karnovsky, M.J. (1998). Involvement of hydrogen peroxide in the differentiation of clonal HD-11EM cells into osteoclast-like cells. J. Cell Physiol. *176*, 574–587.

Suda, N., Morita, I., Kuroda, T., and Murota, S. (1993). Participation of oxidative stress in the process of osteoclast differentiation. Biochim. Biophys. Acta *1157*, 318–323.

Takito, J., Otsuka, H., Yanagisawa, N., Arai, H., Shiga, M., Inoue, M., Nonaka, N., and Nakamura, M. (2015). Regulation of osteoclast multinucleation by the actin cytoskeleton signaling network. J. Cell Physiol. *230*, 395–405.

Tatsumi, S., Ishii, K., Amizuka, N., Li, M., Kobayashi, T., Kohno, K., Ito, M., Takeshita, S., and Ikeda, K. (2007). Targeted ablation of osteocytes induces osteoporosis with defective mechanotransduction. Cell Metab. *5*, 464–475.

Tautzenberger, A., Fortsch, C., Zwerger, C., Dmochewitz, L., Kreja, L., Ignatius, A., and Barth, H. (2013). C3 rho-inhibitor for targeted pharmacological manipulation of osteoclast-like cells. PLoS One *8*, e85695.

Teti, A., Marchisio, P.C., and Zallone, A.Z. (1991). Clear zone in osteoclast function: role of podosomes in regulation of bone-resorbing activity. Am. J. Physiol. *261*, C1–C7.

Touaitahuata, H., Blangy, A., and Vives, V. (2014). Modulation of osteoclast differentiation and bone resorption by Rho GTPases. Small GTPases *5*, e28119.

Tsuda, E., Goto, M., Mochizuki, S., Yano, K., Kobayashi, F., Morinaga, T., and Higashio, K. (1997). Isolation of a novel cytokine from human fibroblasts that specifically inhibits osteoclastogenesis. Biochem. Biophys. Res. Commun. *234*, 137–142.

Udagawa, N., Takahashi, N., Akatsu, T., Tanaka, H., Sasaki, T., Nishihara, T., Koga, T., Martin, T.J., and Suda, T. (1990). Origin of osteoclasts: mature monocytes and macrophages are capable of differentiating into osteoclasts under a suitable microenvironment prepared by bone marrow-derived stromal cells. Proc. Natl. Acad. Sci. USA *87*, 7260–7264.

Usui, M., Xing, L., Drissi, H., Zuscik, M., O'Keefe, R., Chen, D., and Boyce, B.F. (2008). Murine and chicken chondrocytes regulate osteoclastogenesis by producing RANKL in response to BMP2. J. Bone Miner. Res. *23*, 314–325.

van Helden, S.F., Anthony, E.C., Dee, R., and Hordijk, P.L. (2012). Rho GTPase expression in human myeloid cells. PLoS One *7*, e42563.

Varoga, D., Drescher, W., Pufe, M., Groth, G., and Pufe, T. (2009). Differential expression of vascular endothelial growth factor in glucocorticoid-related osteonecrosis of the femoral head. Clin. Orthop. Relat. Res. *467*, 3273–3282.

Verborgt, O., Gibson, G.J., and Schaffler, M.B. (2000). Loss of osteocyte integrity in association with microdamage and bone remodeling after fatigue *in vivo*. J. Bone Miner. Res. *15*, 60–67.

Vignery, A. (2011). Macrophage fusion: the making of a new cell. In: Larsson, L.-I., ed., *Cell Fusions: Regulation and Control*, Springer, New York, pp. 219–231.

Vives, V., Cres, G., Richard, C., Busson, M., Ferrandez, Y., Planson, A.G., Zeghouf, M., Cherfils, J., Malaval, L., and Blangy, A. (2015). Pharmacological inhibition of Dock5 prevents osteolysis by affecting osteoclast podosome organization while preserving bone formation. Nat. Commun. *6*, 6218.

Vives, V., Laurin, M., Cres, G., Larrousse, P., Morichaud, Z., Noel, D., Cote, J.F., and Blangy, A. (2011). The Rac1 exchange factor Dock5 is essential for bone resorption by osteoclasts. J. Bone Miner. Res. *26*, 1099–1110.

Wang, Y., Brooks, P.J., Jang, J.J., Silver, A.S., Arora, P.D., McCulloch, C.A., and Glogauer, M. (2015). Role of actin filaments in fusopod formation and osteoclastogenesis. Biochim. Biophys. Acta *1853*, 1715–1724.

Wang, Y., Lebowitz, D., Sun, C., Thang, H., Grynpas, M.D., and Glogauer, M. (2008). Identifying the relative contributions of Rac1 and Rac2 to osteoclastogenesis. J. Bone Miner. Res. *23*, 260–270.

Weernink, P.A., Meletiadis, K., Hommeltenberg, S., Hinz, M., Ishihara, H., Schmidt, M., and Jakobs, K.H. (2004). Activation of type I phosphatidylinositol 4-phosphate 5-kinase isoforms by the Rho GTPases, RhoA, Rac1, and Cdc42. J. Biol. Chem. *279*, 7840–7849.

Yan, J., Chen, S., Zhang, Y., Li, X., Li, Y., Wu, X., Yuan, J., Robling, A.G., Kapur, R., Chan, R.J., *et al.* (2008). Rac1 mediates the osteoclast gains-in-function induced by haploinsufficiency of Nf1. Hum. Mol. Genet. *17*, 936–948.

Yang, F.C., Atkinson, S.J., Gu, Y., Borneo, J.B., Roberts, A.W., Zheng, Y., Pennington, J., and Williams, D.A. (2001). Rac and Cdc42 GTPases control hematopoietic stem cell shape, adhesion, migration, and mobilization. Proc. Natl. Acad. Sci. USA 98, 5614–5618.

Yang, S., Zhang, Y., Ries, W., and Key, L. (2004). Expression of Nox4 in osteoclasts. J. Cell Biochem. 92, 238–248.

Yavropoulou, M.P. and Yovos, J.G. (2008). Osteoclastogenesis — current knowledge and future perspectives. J. Musculoskelet. Neuronal Interact. 8, 204–216.

Ye, S., Fujiwara, T., Zhou, J., Varughese, K.I., and Zhao, H. (2016). LIS1 regulates osteoclastogenesis through modulation of M-SCF and RANKL signaling pathways and CDC42. Int. J. Biol. Sci. 12, 1488–1499.

Zhang, D., Udagawa, N., Nakamura, I., Murakami, H., Saito, S., Yamasaki, K., Shibasaki, Y., Morii, N., Narumiya, S., Takahashi, N., et al. (1995). The small GTP-binding protein, rho p21, is involved in bone resorption by regulating cytoskeletal organization in osteoclasts. J. Cell Sci. 108 (Pt 6), 2285–2292.

Zhou, J., Ye, S., Fujiwara, T., Manolagas, S.C., and Zhao, H. (2013). Steap4 plays a critical role in osteoclastogenesis in vitro by regulating cellular iron/reactive oxygen species (ROS) levels and cAMP response element-binding protein (CREB) activation. J. Biol. Chem. 288, 30064–30074.

Zhu, T., Chappel, J.C., Hsu, F.F., Turk, J., Aurora, R., Hyrc, K., De Camilli, P., Broekelmann, T.J., Mecham, R.P., Teitelbaum, S.L., et al. (2013). Type I phosphotidylinosotol 4-phosphate 5-kinase gamma regulates osteoclasts in a bifunctional manner. J. Biol. Chem. 288, 5268–5277.

Rhogefs as therapeutic targets 9

Anne Blangy

Centre de Recherche en Biologie cellulaire de Montpellier (CRBM), CNRS-UMR5237, Université de Montpellier, 1919 route de Mende Montpellier 34293, Cedex 05, France

anne.blangy@crbm.cnrs.fr

Keywords: Rho GTPase; exchange factor; inhibitor; aptamer; therapeutic target; osteoclast; osteoporosis; yeast exchange assay.

9.1. Introduction

9.1.1. *The relevance Rho GTPase exchange factors as therapeutic targets*

Rho GTPase signaling pathways are major regulators of eukaryotic cell dynamics, which control normal and pathological processes (Cook *et al.*, 2014). They participate in cell migration, morphology, polarity, and differentiation during embryonic development (Duquette and Lamarche-Vane, 2014; Fort and Théveneau, 2014). They are also involved in the pathological mechanisms of a variety of diseases, including hypertension (Shimokawa *et al.*, 2016), cancer (Lin and Zheng, 2015), and neurodegenerative diseases (Stankiewicz and Linseman, 2014).

Mammals have 20 Rho GTPases (Boureux *et al.*, 2007) and 82 Rho GTPase exchange factors (RhoGEFs) that distribute between two families: the Dbl-related and the Dock-related RhoGEFs (see Chapter 3 by Amin and Ahmadian in this volume). The Dbl family counts 71 members (Jaiswal *et al.*, 2013; Cook *et al.*, 2014), and there are 11 proteins in the Dock family (Gadea and Blangy, 2014). They

161

activate Rho GTPase via their catalytic domain called the DH domain for the Dbl family or the DHR2 domain for the Dock family. The RhoGEFs are multidomain proteins: their catalytic domain is accompanied by various functional domains that can mediate the association of the GEF with membrane receptors or lipids, for instance, or provide the GEF with other enzymatic activities, such as kinase, phosphatase, or even GEF or GAP function toward other Ras-like GTPases.

Rho GTPase signaling pathways can be targeted at various levels: not only the GTPases themselves but also GEFs, GAPs, and downstream effectors. Drugs have been developed against downstream effectors of Rho GTPases, such as Y-27632 that inhibits the kinase Rock (Uehata *et al.*, 1997) or IPA3 that targets the Pak kinases (Deacon *et al.*, 2008). Rock inhibitor Fasudil is used in clinics to modulate pulmonary hypertension and Ripasudil for the treatment of glaucoma (Defert and Boland, 2017). Several inhibitors of Rho GTPases exist naturally in bacteria, such as *Clostridium botulinum* C3 exoenzyme that targets RhoA, RhoB, and RhoC (Sekine *et al.*, 1989). Others were developed for scientific applications, such as Rac inhibitors EHT1864, NSC23766 and its derivative EHop-016 (Gao *et al.*, 2004; Montalvo-Ortiz *et al.*, 2012; Shutes *et al.*, 2007) and Cdc42 inhibitor ML141 (Surviladze *et al.*, 2010) and RhoA inhibitor Rhosin (Shang *et al.*, 2012). But the ubiquitous expression of most Rho GTPases and their implication in fundamental cellular processes do not make Rho GTPase inhibitors suitable for therapeutic applications. In fact, the knockdown of many Rho GTPases is deleterious; for instance, RhoA-null (Pedersen and Brakebusch, 2012), Rac1-null (Sugihara *et al.*, 1998), and Cdc42-null (Chen *et al.*, 2000) mice die early during embryonic development.

RhoGEFs activate their target Rho GTPases in response to different signals, usually transmitted from the extracellular medium by membrane receptors. This results in local and temporal regulation of the activation of Rho GTPases. The majority of RhoGEFs have a restricted tissue and/or subcellular distribution, and they are specific for one Rho GTPase (Cook *et al.*, 2014; Gadea and Blangy, 2014). In pathological context, a number of RhoGEFs were found overexpressed, including Dbl, Vav1/2/3, Ect2, Tiam1/2, P-Rex1/2 in cancer, or bearing activating mutations, such as LARG, BCR (Lin and Zheng, 2015) in cancer, Vav1 in multiple sclerosis (Jagodic *et al.*, 2009), Dock2 in Alzheimer's disease (Cimino *et al.*, 2013), and Dock3 in muscular dystrophy (Alexander *et al.*, 2014). Conversely, there are only sporadic examples of mutation or overexpression described for Rho GTPase, unlike Ras GTPases that are often found mutated in cancers. Thus, RhoGEFs are attractive targets to optimize efficacy and specificity of Rho GTPase signaling inhibition. Several RhoGEFs already qualify as relevant therapeutic targets, and different types of inhibitors have been developed through a variety of strategies that are described in this chapter (Table 1).

Table 1: Inhibitors of Rho GTPase exchange factors.

Target GEF	Inhibitor	Validation	Pathologies	References
		Peptide inhibitor		
TRIO (D2), Tgat	TRIPα, TRIP[E32G]	Cell-free assay, Cellular expression	T-Cell Leukemia	Bouquier et al. (2009a) Schmidt et al. (2002)
DOCK2	DCpep-4-NH2	Cell-free assay, Cell culture	Immune disorders, Alzheimer's disease	Sakamoto et al. (2017)
		RNA inhibitor		
TIAM1	K91	Cell-free assay	Various cancers	Niebel et al. (2013)
		Chemical inhibitor		
TRIO (D1)	ITX3	Cell-free assay, Cell culture	Glioblastoma, Breast cancer	Blangy et al. (2006) Bouquier et al. (2009b)
DOCK5	C21[a]	Cell-free assay, Cell culture, Mouse models of pathologies	Osteolytic diseases, Osteoporosis, Bone metastases, Inflammatory diseases	Vives et al. (2011, 2015)
DOCK2	CPYPP	Cell-free assay, Cell culture, Mouse (T-cell homing)	Immune related disorders, Alzheimer's disease	Nishikimi et al. (2012)
ARHHEF12/LARG	Y16	Cell-free assay, Cell culture	Acute myeloid Leukemia	Shang et al. (2013)
AKAP13/LBC	A13	Cell-free assay, Cell culture	Various cancers	Diviani et al. (2016)

Note: [a] C21, the inhibitor of Dock5 (CAS 54129-15-6), should not be mistaken for the inhibitor of the protein arginine methyltransferase PRMT1 (CAS 1229236-78) and for the nonpeptide selective AT2 receptor agonist M24 (CAS 477775-14-7), which were also named C21.

9.1.2. *Targeting RhoGEF activity*

The activation of a GTPase by an exchange factor is a complex enzymatic reaction. The precise molecular mechanism driving the catalysis of nucleotide release by the GTPase is different between Dbl- (Rossman *et al.*, 2002) and Dock-related RhoGEFs (Yang *et al.*, 2009). But overall, the sequence of interactions between the GTPase and the GEF is similar. The first step is the formation of a complex between the GEF and the inactive GDP-bound GTPase. In this low-affinity complex, the exchange factor provokes conformational modifications in the GTPase causing the release of the guanine nucleotide. This results in a stable complex between the nucleotide-free GTPase and the GEF. GTP destabilizes this complex and then binds to the empty nucleotide pocket of the GTPase, provoking the release of the active GTP-bound GTPase from the GEF.

To inhibit the exchange reaction, the strategy is to target the catalytic DH or DHR2 domain of the GEF. Aiming more specifically at the interface between RhoGEF and the GTPase during the nucleotide exchange reaction can render the inhibition even more specific. This approach is made feasible with the increasing number of Rho GTPase–GEF complexes available in databases (http://www.rcsb.org/pdb). The approach used so far to obtain RhoGEF inhibitors has been to prevent the interaction between the RhoGEF and the GTPase. This is the strategy used in nature by enteric bacterial pathogens that produce type III effector EspH, a small 20 kDa protein that interacts with various GEFs for RhoA, preventing their binding to the Rho GTPase (Dong *et al.*, 2010). Another solution, which could be considered for the future design of RhoGEF inhibitors, is to stabilize an intermediate step of the exchange reaction, thereby freezing the GEF–GTPase complex and compromising the activation of the GTPase. This is again a strategy efficiently developed in nature by fungi; they produce the macrocyclic lactone Brefeldin A, inhibit the activation of ARF-family GTPases by locking the complex between the GTPase and the GEF (Peyroche *et al.*, 1999).

9.2. Different types of RhoGEF inhibitors

9.2.1. *Peptides and nucleic acids*

9.2.1.1. *Peptidic inhibitors*

Historically, the first inhibitor of a RhoGEF is Trio inhibitory peptide α (TRIPα), which targets the exchange factor Trio. Trio is an unusual exchange factor in that it exhibits two exchange domains (Debant *et al.*, 1996): Trio-D1 can activate the GTPases Rac1 (Debant *et al.*, 1996) and RhoG (Blangy *et al.*, 2000), and Trio-D2 activates the GTPase RhoA (Debant *et al.*, 1996). The *Trio* gene is expressed as several splice variants including Tgat, an oncogenic form of Trio identified in adult T-cell leukemia (ATL) patients. Tgat only has the Trio-D2 exchange domain, responsible for its oncogenic activity (Yoshizuka *et al.*, 2004).

The strategy used to find inhibitors of Trio-D2 was to identify peptides able to bind the GEF using the yeast two-hybrid system. The screening was performed on a library of 2×10^6 plasmids expressing random 20-aminoacid peptides (peptide aptamers), whose conformation is constrained by fusion to the bacterial protein Thioredoxin A (Colas *et al.*, 1996). The expression of a peptide binding to Trio-D2 translates into the ability of yeast to grow in appropriate selective medium (Schmidt *et al.*, 2002). As the screening is performed in a living organism, toxic aptamers result eliminated, as it prevents yeast growth. This system allows easy testing of aptamer selectivity for the GEF of interest, by monitoring its ability to bind other GEFs. This way, TRIPα was selected as it was able to bind to Trio-D2 but not to other GEFs for RhoA, such as MCF2/Dbl, ARHGEF1/p115-RhoGEF, and ARHGEF11/PDZ-RhoGEF (Schmidt *et al.*, 2002). Mutagenesis can be performed in the aptamer sequence to identify important residues and increase the interaction potential. Thereby, mutations were selected in TRIPα that increase its capacity of binding the GEF. Still, binding does not mean inhibition and the effect of the peptide on the exchange reaction must be tested. This can be done either in a cell-free exchange assay, where the GTPase, the GEF, and the peptide aptamer are combined as purified proteins, or in a cellular system in which the activation of the GTPase in response to the expression of the GEF, with and without the aptamer, is monitored by pull-down assays that detect the active GTP-bound GTPase. In both systems, TRIPα proved an efficient inhibitor of Trio-D2 (Blangy *et al.*, 2006; Schmidt *et al.*, 2002), and more potent inhibitors were derived by mutagenesis of TRIPα. When expressed in NIH-3T3 cell transformed with Tgat, TRIPα derivative TRIP[E32G] was able to diminish the activation of RhoA and Tgat oncogenic potential; TRIP[E32G] expression also reduced the growth of Tgat-transformed cells after their subcutaneous engraftment in nude mice (Bouquier *et al.*, 2009a).

Very recently, a small peptide was developed to inhibit Dock2 (Sakamoto *et al.*, 2017), an exchange factor for Rac GTPases. Dock2 controls lymphocyte activation and migration, and it is considered an interesting target in the context of immune-related disorders (Gadea and Blangy, 2014) and also in Alzheimer's disease (Cimino *et al.*, 2013). The identification of peptides able to interact with Dock2 was done by phage display. Random peptides displayed by the T7 phage were selected for their ability to bind to the DHR2 exchange domain of Dock2 immobilized on beads and then to be displaced by Rac1. Several rounds of amplification selected a phage expressing the 17-aminoacid peptide DCpep-4-NH2 (LNRCVAKYHGYPW-

CRRR). DCpep-4-NH2 inhibited the ability of Dock2 to bind and activate Rac1 in a cell-free assay. Conversely, it did not affect the interaction of Rac1 with Dock1, a GEF closely related to Dock2. DCpep-4-NH2, fused or not to a cell-penetrating peptide, and added to the culture medium, hindered sphingosine-1-phosphate-induced lymphocytic cell migration, a process known to rely on Dock2 (Sakamoto *et al.*, 2017).

9.2.1.2. *RNA inhibitors*

Small nucleic acid molecules (nucleic acid aptamers) are able to bind proteins with high affinity and specificity. Methods to develop such molecules as targets of various types of proteins using the systematic evolution of ligands by exponential enrichment (SELEX) have been expanding for the last three decades (Mallikaratchy, 2017). The first RNA aptamer targeting the activity of an exchange factor was M69, which is active towards cytohesins, a family of GEF for ARF-type small GTPases (Mayer *et al.*, 2001). Some 12 years later, the first RNA aptamer inhibitor of a RhoGEF was engineered (Niebel *et al.*, 2013): K91 is an inhibitor of the Rac GEF Tiam1, T-lymphoma invasive, and metastasis-inducing protein 1, which involved in cancer (Boissier and Huynh-Do, 2014). The approach was to select for RNAs able to bind *in vitro* to the purified exchange domain of Tiam1 among a library of 4×10^{14} random 50-nucleotide aptamer RNAs constrained within a constant 40-nucleotide RNA sequence. Sixteen rounds of selection followed by PCR amplification led to the identification of 33 RNA aptamers with high binding affinity for Tiam1-exchange domain (Niebel *et al.*, 2013). Similar to peptide aptamers, the ability of the RNA aptamer to inhibit the exchange reaction must be confirmed as the selection procedure is only based on its ability to bind the target GEF. RNA aptamer K91 was found to inhibit the activation of Rac1 by Tiam1 *in vitro*, but its ability to inhibit Tiam1 in cellular systems remains to be confirmed (Niebel *et al.*, 2013).

RNA and peptide aptamers proved efficient at inhibiting RhoGEF activity, and they are easily amenable to optimization by mutagenesis. They can be expressed in cell to monitor their biological effects and test their toxicity. Still they suffer an important limitation regarding therapeutic usage. In fact, RhoGEFs are intracellular targets, and the difficult challenge to deliver RNAs and peptides into a cell within a living organism remains a big limitation to their utilization as therapeutic agents.

9.2.2. *Chemical compounds*

Thus far, small chemical compounds represent the vast majority of therapeutic agents. Several molecules were developed to target RhoGEFs through various approaches (Table 1).

9.2.2.1. *Functional approach: the yeast exchange assay*

The first inhibitor of a RhoGEF was designed to target Trio-D1 (Blangy *et al.*, 2006), one of the two DH domain of Trio that is specific for Rac1 (Debant *et al.*, 1996) and RhoG (Blangy *et al.*, 2000). This inhibitor was identified taking advantage of the yeast exchange assay, a method developed in live yeast to monitor the activation of a Rho GTPase by an exchange factor (De Toledo *et al.*, 2000). In this reporter assay, the expression of Trio-D1 induced the activation of RhoG resulting in its binding to its effector kinectin. This was monitored by the expression of β-galactosidase and by yeast becoming auxotrophic for histidine (Blangy *et al.*, 2006). In this experimental setup, a chemical library of 2,640 compounds was screened for species able to inhibit yeast growth in histidine-deprived medium, as indicative of the inhibition of RhoG activation by Trio-D1. In parallel, the same library was screened in growth medium supplemented with histidine, to eliminate any molecule that would inhibit yeast growth, indicative of cytotoxicity. Thereby, a series of compounds were identified as the first chemical inhibitors of an RhoGEF. Among these, three molecules displayed selectivity: they efficiently inhibited Trio-D1 but neither Arhgef17, a GEF for RhoA, nor cytohesin-2, a GEF for the small GTPase Arf1 (Blangy *et al.*, 2006). Toxicity assays on mammalian cells restricted usable TRIO-D1 inhibitors to the compound Inhibitor of Trio eXchange 1 (ITX1): 2-(5-chloro-2-ethoxybenzylidene) [1,3] thiazolo [3,2-a] benzimidazol-3(2H)-one (Bouquier *et al.*, 2009b). Analogs of ITX1 were tested, and molecule ITX3 (CAS 347323-96-0) proved efficient at inhibiting Trio-D1 in cell-free assays and also a variety of Trio-dependent cellular functions, ranging from myogenic differentiation (Bouquier *et al.*, 2009b) to leukocyte transendothelial migration (van Rijssel *et al.*, 2012) and endothelial barrier formation (Timmerman *et al.*, 2015).

    The yeast exchange assay was also used to identify the first chemical inhibitor of a RhoGEF from the Dock family. C21 (N-(3,5- dichlorophenyl) benzene-sulfonamide, CAS 54129-15-6) was characterized as an inhibitor of Dock5, an exchange factor for Rac (Vives *et al.*, 2011). C21 can inhibit Dock5 in cell-free assays, in culture cells, and *in vivo* in the mouse (Vives *et al.*, 2011, 2015) (see Section 9.3).

9.2.2.2. *Cell-free protein–protein interaction*

Another strategy to find inhibitors of exchange factors is to look for molecules that can disrupt the interaction between the GEF and the GTPase. A library of 9,392 chemical compounds was screened for molecules able to prevent the binding of Rac1 to the immobilized DHR2-exchange domain of Dock2.

This cell-free approach picked up CPYPP (CAS 310460-39-0), which inhibits the activation of Rac1 by Dock2 in cell-free assays as well as in HEK293T cells. CPYPP also prevents lymphocyte migration and leukocyte activation in culture, two processes relying on Rac activation by Dock2. In the mouse, intraperitoneal injection of CPYPP (250 mg/kg), just 1 hour before the adoptive transfer of spleen cells, was found to prevent T-cell homing to the lymph node and to the spleen (Nishikimi et al., 2012). This was the first example of a RhoGEF inhibitor proven to be active in a whole organism.

9.2.2.3. Rational design

Rather than performing a physical screening, the rational design is based on computational modeling. From the crystal or modeled structure of the target, virtual molecules are screened or designed according to their capacity to dock onto the target, which is likely to interfere with the pathway of interest. Rational design proved successful for the development of inhibitors of Rho GTPases including Rac1 (NSC23766; Gao et al., 2004) and RhoA (Rhosin; Shang et al., 2012). Several crystal structures of RhoGEF·Rho GTPase complexes were solved in the recent years, which allow following the same strategy to develop molecules that target the RhoGEF instead of the GTPase. In fact, the RhoGEF·Rho GTPase cocrystal unravels the exact binding regions between the two proteins and highlights the interactions between individual amino acids of each protein that are necessary to form the complex and/or important for the nucleotide-exchange reaction (Snyder et al., 2002; Yang et al., 2009). Thereby, it is possible to highlight small pockets in the RhoGEF where the binding of a small chemical compound is likely to interfere with the formation of the RhoGEF·Rho GTPase complex.

LARG/ARHGEF12, p115-RhoGEF/ARHGEF1, and PDZ-RhoGEF/ARHGEF11 are GEFs for RhoA; they are RGS-RhoGEFs meaning that they are regulated by heterotrimeric G-proteins (Chikumi et al., 2004). These GEFs are relevant therapeutic targets in the context of various pathologies including cancer (Reuther et al., 2001), lung hypertension (Guilluy et al., 2010) and obesity (Chang et al., 2015). The crystal structure available for LARG exchange domain in complex with RhoA (PDB 1X86) was used to identify interaction sites between the GEF and the GTPase. This highlighted a concave region in LARG, between residues Asn975 and Arg986, into which RhoA sends a protrusion. The structure of this small domain was used for the screening in silico of 4 million compounds in the ZINC library of virtual compounds, to identify molecules able to dock into the groove. The 49 best hits were synthetized and further validated for their capacity to affect the interaction between RhoA and LARG. This rational drug design strategy identified Y16 (CAS 429653-73-6) as able to prevent the interaction between LARG and RhoA (Shang *et al.*, 2013). Y16 possesses a good selectivity as it affects the binding of RhoA to LARG, p115-RhoGEF, and PDZ-RhoGEF but not to the closely related RhoA GEFs Dbl/MCF2L and Lbc/AKAP13. Y16 has no effect on the binding of Rac1 to Tiam1 and of Cdc42 to its GEF Intersectin-1. In culture cells, Y16 blocks stress fiber formation in response to lysophosphatidic acid, a process driven by RhoA activation downstream of heterotrimeric G-proteins. Interestingly, Y16 and the RhoA-inhibitor Rhosin display a synergistic effect to block the proliferation and invasion of breast cancer cells (Shang *et al.*, 2013).

Y16 does not bind to Lbc, another RGS-RhoGEF closely related to LARG. A model for the RhoA–Lbc complex was built by structural analogy after the RhoA–LARG complex and the amino acids involved in the interaction between the RhoA and Lbc were deduced (Diviani *et al.*, 2016). A virtual screening on the ZINC database highlighted 30 compounds likely to interfere with the formation of the RhoA–Lbc complex. They were synthetized and tested for their ability to block the interaction between RhoA and Lbc in a cellular system, which selected molecule A13 (4-[(4Z)-3-methyl-5-oxo-4-[[5-[3(trifluoromethyl) phenyl] furan-2-yl] methylidene] pyrazol-1-yl] benzoate). A13 affected the activation of RhoA induced by Lbc expression in 293T cells. A13 successfully inhibited the cellular effect driven by RhoA activation by Lbc, in particular NIH-3T3 cell transformation. Interestingly, A13 and Y16 display distinct selectivity for RhoA GEFs. A13 was found to block the binding of RhoA to LARG and PDZ-RhoGEF but not to p115RhoGEF. A13 also interfered with other RhoA GEFs: p114RhoGEF/ARHGEF18, p190RhoGEF/ARHGEF28, and GEF-H1/ARHGEF2 but not p63RhoGEF/ARHGEF25 and Net1 (Diviani *et al.*, 2016).

Hopefully, *in vivo* assays will soon confirm the potentiality of Y16 and A13 RhoGEF inhibitors in the context of cancer.

9.3. An example of preclinical application of a RhoGEF inhibitor: Dock5 and osteolytic diseases

9.3.1. *Dock5 is necessary for bone resorption by osteoclasts*

Osteoclasts are essential for the maintenance of the skeleton. They degrade old or damaged bone, and osteoblasts replace it with new bone. The balanced activity of osteoclasts and osteoblasts throughout life is essential to maintain the health of the skeleton and its adaptation to loading constraints. But a variety of physiological and pathological situations exacerbate osteoclast activity, causing an excess of bone resorption over formation. This leads to progressive bone loss, osteoporosis and bone frailty. This occurs upon sexual hormone decay, for instance, after menopause (Frenkel *et al.*, 2010), in inflammatory diseases such as rheumatoid arthritis (Redlich and Smolen, 2012) and in bone metastasis, in particular of breast cancer (Weilbaecher *et al.*, 2011). Increased osteoclast activity is also an iatrogenic effect of various medical treatments including corticosteroids (Canalis *et al.*, 2007) and cancer chemotherapy (Drake, 2013). In these situations, medications to inhibit osteoclast activity are often associated with the frontline treatment of the disease, to prevent osteoporosis and reduce the risk of fractures, pain, and disability.

Dock5 was identified as an activator of Rac1 essential for bone resorption by osteoclasts in culture and *in vivo* in the mouse (Vives *et al.*, 2011). The activation of Rac1 by Dock5 participates in the organization of osteoclast adhesion structures into a belt of podosomes to form the architecture of the bone-resorption apparatus (Touaitahuata *et al.*, 2014a). In the absence of Dock5, osteoclasts adhere on the bone, but they fail to degrade it. In the mouse, the genetic deletion of Dock5 expectedly results in increased bone mass, while animals grow and behave normally and they remain fertile (Touaitahuata *et al.*, 2014b; Vives *et al.*, 2011). Therefore, Dock5 appears an attractive target in the context of osteolytic diseases to control the excess of bone resorption by osteoclasts.

9.3.2. *An inhibitor of Dock5 can prevent pathological bone loss*

A small molecule hindering the activation of Rac1 by Dock5 was identified using the yeast exchange assay (Blangy *et al.*, 2006) and validated further using biochemistry and cell-culture assays (Vives *et al.*, 2011, 2015). In particular, the molecule C21 prevented Rac1 activation by Dock5 in cell-free and reporter cell assays. In osteoclast in culture, C21 reduced the activity of Rac1, disorganized the belt of podosomes, and hindered their capacity to resorb the bone. *In vivo* in the mouse, daily injections of C21 up to 25 mg/kg during 1 month did not provoke any measurable side effects on the behavior, the blood cell counts, and the weight of the mice; it caused no toxic effect to the liver and the kidneys. C21 was tested for its ability to protect against pathological bone loss in mouse models of human osteolytic diseases: sexual hormone deficiency, inflammation, and bone metastases. In these three disease models, the systemic administration of C21 during 1 month efficiently protected the animals against pathological bone loss (Vives *et al.*, 2015). Therefore, targeting Rac1 activation by Dock5 appears as a beneficial and feasible strategy in the context of osteolytic diseases.

This study is a proof of concept that RhoGEF can indeed constitute relevant therapeutic targets and that they are amenable for long-term inhibition in the context of a whole organism.

Acknowledgments

This work was supported by fundings from, CNRS, Montpellier University and the Fondation pour la Recherche Médicale (Equipe FRM DEQ20160334933).

References

Alexander, M.S., Casar, J.C., Motohashi, N., Vieira, N.M., Eisenberg, I., Marshall, J.L., Gasperini, M.J., Lek, A., Myers, J.A., Estrella, E.A., et al. (2014). MicroRNA-486-dependent modulation of DOCK3/PTEN/AKT signaling pathways improves muscular dystrophy-associated symptoms. J. Clin. Invest. 124, 2651–2667.

Blangy, A., Vignal, E., Schmidt, S., Debant, A., Gauthier-Rouviere, C., and Fort, P. (2000). TrioGEF1 controls Rac- and Cdc42-dependent cell structures through the direct activation of rhoG. J. Cell Sci. 113 (Pt 4), 729–739.

Blangy, A., Bouquier, N., Gauthier-Rouvière, C., Schmidt, S., Debant, A., Leonetti, J.-P., and Fort, P. (2006). Identification of TRIO-GEFD1 chemical inhibitors using the yeast exchange assay. Biol. Cell 98, 511–522.

Boissier, P., and Huynh-Do, U. (2014). The guanine nucleotide exchange factor Tiam1: a Janus-faced molecule in cellular signaling. Cell. Signal. 26, 483–491.

Bouquier, N., Fromont, S., Zeeh, J.-C., Auziol, C., Larrousse, P., Robert, B., Zeghouf, M., Cherfils, J., Debant, A., and Schmidt, S. (2009a). Aptamer-derived peptides as potent inhibitors of the oncogenic RhoGEF Tgat. Chem. Biol. 16, 391–400.

Bouquier, N., Vignal, E., Charrasse, S., Weill, M., Schmidt, S., Léonetti, J.-P., Blangy, A., and Fort, P. (2009b). A cell active chemical GEF inhibitor selectively targets the Trio/RhoG/Rac1 signaling pathway. Chem. Biol. 16, 657–666.

Boureux, A., Vignal, E., Faure, S., and Fort, P. (2007). Evolution of the Rho family of ras-like GTPases in eukaryotes. Mol. Biol. Evol. 24, 203–216.

Canalis, E., Mazziotti, G., Giustina, A., and Bilezikian, J.P. (2007). Glucocorticoid-induced osteoporosis: pathophysiology and therapy. Osteoporos. Int. J. Establ. Result Coop. Eur. Found. Osteoporos. Natl. Osteoporos. Found. USA 18, 1319–1328.

Chang, Y.-J., Pownall, S., Jensen, T.E., Mouaaz, S., Foltz, W., Zhou, L., Liadis, N., Woo, M., Hao, Z., Dutt, P., et al. (2015). The Rho-guanine nucleotide exchange factor PDZ-RhoGEF governs susceptibility to diet-induced obesity and type 2 diabetes. eLife 4, e06011.

Chen, F., Ma, L., Parrini, M.C., Mao, X., Lopez, M., Wu, C., Marks, P.W., Davidson, L., Kwiatkowski, D.J., Kirchhausen, T., et al. (2000). Cdc42 is required for PIP(2)-induced actin polymerization and early development but not for cell viability. Curr. Biol. *10*, 758–765.

Chikumi, H., Barac, A., Behbahani, B., Gao, Y., Teramoto, H., Zheng, Y., and Gutkind, J.S. (2004). Homo- and hetero-oligomerization of PDZ-RhoGEF, LARG and p115RhoGEF by their C-terminal region regulates their in vivo Rho GEF activity and transforming potential. Oncogene *23*, 233–240.

Cimino, P.J., Yang, Y., Li, X., Hemingway, J.F., Cherne, M.K., Khademi, S.B., Fukui, Y., Montine, K.S., Montine, T.J., and Keene, C.D. (2013). Ablation of the microglial protein DOCK2 reduces amyloid burden in a mouse model of Alzheimer's disease. Exp. Mol. Pathol. *94*, 366–371.

Colas, P., Cohen, B., Jessen, T., Grishina, I., McCoy, J., and Brent, R. (1996). Genetic selection of peptide aptamers that recognize and inhibit cyclin-dependent kinase 2. Nature *380*, 548–550.

Cook, D.R., Rossman, K.L., and Der, C.J. (2014). Rho guanine nucleotide exchange factors: regulators of Rho GTPase activity in development and disease. Oncogene *33*, 4021–4035.

De Toledo, M., Colombo, K., Nagase, T., Ohara, O., Fort, P., and Blangy, A. (2000). The yeast exchange assay, a new complementary method to screen for Dbl-like protein specificity: identification of a novel RhoA exchange factor. FEBS Lett. *480*, 287–292.

Deacon, S.W., Beeser, A., Fukui, J.A., Rennefahrt, U.E., Myers, C., Chernoff, J., and Peterson, J.R. (2008). An isoform-selective, small-molecule inhibitor targets the autoregulatory mechanism of p21-activated kinase. Chem. Biol. *15*, 322–331.

Debant, A., Serra-Pages, C., Seipel, K., O'Brien, S., Tang, M., Park, S.H., and Streuli, M. (1996). The multidomain protein Trio binds the LAR transmembrane tyrosine phosphatase, contains a protein kinase domain, and has separate rac-specific and rho-specific guanine nucleotide exchange factor domains. Proc. Natl. Acad. Sci. USA *93*, 5466–5471.

Defert, O. and Boland, S. (2017). Rho kinase inhibitors: a patent review (2014–2016). Expert Opin. Ther. Pat. *27*, 507–515.

Diviani, D., Raimondi, F., Del Vescovo, C.D., Dreyer, E., Reggi, E., Osman, H., Ruggieri, L., Gonano, C., Cavin, S., Box, C.L., et al. (2016). Small-molecule protein-protein interaction inhibitor of oncogenic Rho signaling. Cell Chem. Biol. *23*, 1135–1146.

Dong, N., Liu, L., and Shao, F. (2010). A bacterial effector targets host DH-PH domain RhoGEFs and antagonizes macrophage phagocytosis. EMBO J. *29*, 1363–1376.

Drake, M.T. (2013). Osteoporosis and cancer. Curr. Osteoporos. Rep. *11*, 163–170.

Duquette, P.M., and Lamarche-Vane, N. (2014). Rho GTPases in embryonic development. Small GTPases *5*, e972857.

Fort, P. and Théveneau, E. (2014). PleiotRHOpic: Rho pathways are essential for all stages of neural crest development. Small GTPases *5*, e27975.

Frenkel, B., Hong, A., Baniwal, S.K., Coetzee, G.A., Ohlsson, C., Khalid, O., and Gabet, Y. (2010). Regulation of adult bone turnover by sex steroids. J. Cell Physiol. *224*, 305–310.

Gadea, G. and Blangy, A. (2014). Dock-family exchange factors in cell migration and disease. Eur. J. Cell Biol. *93*, 466–477.

Gao, Y., Dickerson, J.B., Guo, F., Zheng, J., and Zheng, Y. (2004). Rational design and characterization of a Rac GTPase-specific small molecule inhibitor. Proc. Natl. Acad. Sci. USA *101*, 7618–7623.

Guilluy, C., Brégeon, J., Toumaniantz, G., Rolli-Derkinderen, M., Retailleau, K., Loufrani, L., Henrion, D., Scalbert, E., Bril, A., Torres, R.M., *et al.* (2010). The Rho exchange factor Arhgef1 mediates the effects of angiotensin II on vascular tone and blood pressure. Nat. Med. *16*, 183–190.

Jagodic, M., Colacios, C., Nohra, R., Dejean, A.S., Beyeen, A.D., Khademi, M., Casemayou, A., Lamouroux, L., Duthoit, C., Papapietro, O., *et al.* (2009). A role for VAV1 in experimental autoimmune encephalomyelitis and multiple sclerosis. Sci. Transl. Med. *1*, 10ra21.

Jaiswal, M., Dvorsky, R., and Ahmadian, M.R. (2013). Deciphering the molecular and functional basis of Dbl family proteins: a novel systematic approach toward classification of selective activation of the Rho family proteins. J. Biol. Chem. *288*, 4486–4500.

Lin, Y. and Zheng, Y. (2015). Approaches of targeting Rho GTPases in cancer drug discovery. Expert Opin. Drug Discov. *10*, 991–1010.

Mallikaratchy, P. (2017). Evolution of complex target SELEX to identify aptamers against mammalian cell-surface antigens. Molecules *22*, 215.

Mayer, G., Blind, M., Nagel, W., Bohm, T., Knorr, T., Jackson, C.L., Kolanus, W., and Famulok, M. (2001). Controlling small guanine-nucleotide-exchange factor function through cytoplasmic RNA intramers. Proc. Natl. Acad. Sci. USA *98*, 4961–4965.

Montalvo-Ortiz, B.L., Castillo-Pichardo, L., Hernández, E., Humphries-Bickley, T., De la Mota-Peynado, A., Cubano, L.A., Vlaar, C.P., and Dharmawardhane, S. (2012). Characterization of EHop-016, novel small molecule inhibitor of Rac GTPase. J. Biol. Chem. *287*, 13228–13238.

Niebel, B., Wosnitza, C.I., and Famulok, M. (2013). RNA-aptamers that modulate the RhoGEF activity of Tiam1. Bioorg. Med. Chem. *21*, 6239–6246.

Nishikimi, A., Uruno, T., Duan, X., Cao, Q., Okamura, Y., Saitoh, T., Saito, N., Sakaoka, S., Du, Y., Suenaga, A., *et al.* (2012). Blockade of inflammatory responses by a small-molecule inhibitor of the Rac activator DOCK2. Chem. Biol. *19*, 488–497.

Pedersen, E., and Brakebusch, C. (2012). Rho GTPase function in development: how in vivo models change our view. Exp. Cell Res. *318*, 1779–1787.

Peyroche, A., Antonny, B., Robineau, S., Acker, J., Cherfils, J., and Jackson, C.L. (1999). Brefeldin A acts to stabilize an abortive ARF-GDP-Sec7 domain protein complex: involvement of specific residues of the Sec7 domain. Mol. Cell *3*, 275–285.

Redlich, K. and Smolen, J.S. (2012). Inflammatory bone loss: pathogenesis and therapeutic intervention. Nat. Rev. Drug Discov. *11*, 234–250.

Reuther, G.W., Lambert, Q.T., Booden, M.A., Wennerberg, K., Becknell, B., Marcucci, G., Sondek, J., Caligiuri, M.A., and Der, C.J. (2001). Leukemia-associated Rho guanine nucleotide exchange factor, a Dbl family protein found mutated in leukemia, causes transformation by activation of RhoA. J. Biol. Chem. *276*, 27145–27151.

van Rijssel, J., Kroon, J., Hoogenboezem, M., van Alphen, F.P.J., de Jong, R.J., Kostadinova, E., Geerts, D., Hordijk, P.L., and van Buul, J.D. (2012). The Rho-guanine nucleotide exchange factor Trio controls leukocyte transendothelial migration by promoting docking structure formation. Mol. Biol. Cell 23, 2831–2844.

Rossman, K.L., Worthylake, D.K., Snyder, J.T., Siderovski, D.P., Campbell, S.L., and Sondek, J. (2002). A crystallographic view of interactions between Dbs and Cdc42: PH domain-assisted guanine nucleotide exchange. EMBO J. 21, 1315–1326.

Sakamoto, K., Adachi, Y., Komoike, Y., Kamada, Y., Koyama, R., Fukuda, Y., Kadotani, A., Asami, T., and Sakamoto, J. (2017). Novel DOCK2-selective inhibitory peptide that suppresses B-cell line migration. Biochem. Biophys. Res. Commun. 483, 183–190.

Schmidt, S., Diriong, S., Méry, J., Fabbrizio, E., and Debant, A. (2002). Identification of the first Rho-GEF inhibitor, TRIPalpha, which targets the RhoA-specific GEF domain of Trio. FEBS Lett. 523, 35–42.

Sekine, A., Fujiwara, M., and Narumiya, S. (1989). Asparagine residue in the rho gene product is the modification site for botulinum ADP-ribosyltransferase. J. Biol. Chem. 264, 8602–8605.

Shang, X., Marchioni, F., Sipes, N., Evelyn, C.R., Jerabek-Willemsen, M., Duhr, S., Seibel, W., Wortman, M., and Zheng, Y. (2012). Rational design of small molecule inhibitors targeting RhoA subfamily Rho GTPases. Chem. Biol. 19, 699–710.

Shang, X., Marchioni, F., Evelyn, C.R., Sipes, N., Zhou, X., Seibel, W., Wortman, M., and Zheng, Y. (2013). Small-molecule inhibitors targeting G-protein-coupled Rho guanine nucleotide exchange factors. Proc. Natl. Acad. Sci. USA 110, 3155–3160.

Shimokawa, H., Sunamura, S., and Satoh, K. (2016). RhoA/Rho-kinase in the cardiovascular system. Circ. Res. 118, 352–366.

Shutes, A., Onesto, C., Picard, V., Leblond, B., Schweighoffer, F., and Der, C.J. (2007). Specificity and mechanism of action of EHT 1864, a novel small molecule inhibitor of Rac family small GTPases. J. Biol. Chem. 282, 35666–35678.

Snyder, J.T., Worthylake, D.K., Rossman, K.L., Betts, L., Pruitt, W.M., Siderovski, D.P., Der, C.J., and Sondek, J. (2002). Structural basis for the selective activation of Rho GTPases by Dbl exchange factors. Nat. Struct. Biol. 9, 468–475.

Stankiewicz, T.R. and Linseman, D.A. (2014). Rho family GTPases: key players in neuronal development, neuronal survival, and neurodegeneration. Front. Cell. Neurosci. 8, 314.

Sugihara, K., Nakatsuji, N., Nakamura, K., Nakao, K., Hashimoto, R., Otani, H., Sakagami, H., Kondo, H., Nozawa, S., Aiba, A., et al. (1998). Rac1 is required for the formation of three germ layers during gastrulation. Oncogene 17, 3427–3433.

Surviladze, Z., Waller, A., Strouse, J.J., Bologa, C., Ursu, O., Salas, V., Parkinson, J.F., Phillips, G.K., Romero, E., Wandinger-Ness, A., et al. (2010). A potent and selective inhibitor of Cdc42 GTPase. In: Probe Reports from the NIH Molecular Libraries Program, National Center for Biotechnology Information, Bethesda, MD.

Timmerman, I., Heemskerk, N., Kroon, J., Schaefer, A., van Rijssel, J., Hoogenboezem, M., van Unen, J., Goedhart, J., Gadella, T.W.J., Yin, T., et al. (2015). A local VE-cadherin and Trio-based signaling complex stabilizes endothelial junctions through Rac1. J. Cell Sci. 128, 3041–3054.

Touaitahuata, H., Blangy, A., and Vives, V. (2014a). Modulation of osteoclast differentiation and bone resorption by Rho GTPases. Small GTPases 5, e28119.

Touaitahuata, H., Cres, G., de Rossi, S., Vives, V., and Blangy, A. (2014b). The mineral dissolution function of osteoclasts is dispensable for hypertrophic cartilage degradation during long bone development and growth. Dev. Biol. 393, 57–70.

Uehata, M., Ishizaki, T., Satoh, H., Ono, T., Kawahara, T., Morishita, T., Tamakawa, H., Yamagami, K., Inui, J., Maekawa, M., et al. (1997). Calcium sensitization of smooth muscle mediated by a Rho-associated protein kinase in hypertension. Nature 389, 990–994.

Vives, V., Laurin, M., Cres, G., Larrousse, P., Morichaud, Z., Noel, D., Côté, J.-F., and Blangy, A. (2011). The Rac1 exchange factor Dock5 is essential for bone resorption by osteoclasts. J. Bone Miner. Res. 26, 1099–1110.

Vives, V., Cres, G., Richard, C., Busson, M., Ferrandez, Y., Planson, A.-G., Zeghouf, M., Cherfils, J., Malaval, L., and Blangy, A. (2015). Pharmacological inhibition of Dock5 prevents osteolysis by affecting osteoclast podosome organization while preserving bone formation. Nat. Commun. 6, 6218.

Weilbaecher, K.N., Guise, T.A., and McCauley, L.K. (2011). Cancer to bone: a fatal attraction. Nat. Rev. Cancer 11, 411–425.

Yang, J., Zhang, Z., Roe, S.M., Marshall, C.J., and Barford, D. (2009). Activation of Rho GTPases by DOCK exchange factors is mediated by a nucleotide sensor. Science 325, 1398–1402.

Yoshizuka, N., Moriuchi, R., Mori, T., Yamada, K., Hasegawa, S., Maeda, T., Shimada, T., Yamada, Y., Kamihira, S., Tomonaga, M., et al. (2004). An alternative transcript derived from the Trio locus encodes a guanosine nucleotide exchange factor with mouse cell-transforming potential. J. Biol. Chem. 279, 43998–44004.

Endothelial specific GTPase signaling during leukocyte extravasation

10

Sofia Morsing, Lilian Schimmel, Jos van Rijssel, and Jaap D. van Buul*

Department of Plasma Proteins, Molecular Cell Biology Lab,
Sanquin Research and Landsteiner Laboratory, Academic Medical Center,
University of Amsterdam, Plesmanlaan 125, Amsterdam 1066 CX,
The Netherlands

*j.vanbuul@sanquin.nl

Keywords: Rho GTPase, cytoskeleton, leukocytes, endothelial transmigration, inflammation, VE-cadherin.

10.1. Introduction

Inflammation is part of the complex biological response of body tissues to harmful stimuli, such as pathogens or damaged cells (Pober and Sessa, 2007). It serves as a protective response that involves leukocytes, blood vessels, and molecular mediators with the purpose of eliminating the initial cause of cell injury, clearing out necrotic cells and tissues damaged from the original insult and the inflammatory process, and initiating tissue repair. Too little inflammation leads to progressive tissue destruction by the harmful stimulus (e.g., bacteria) and compromises the survival of the organism. In contrast, chronic inflammation may lead to a host of diseases, such as hay fever, periodontitis, atherosclerosis, rheumatoid arthritis, and even cancer (e.g., gallbladder carcinoma). Inflammation is therefore normally closely regulated by the body. Normal physiology is associated with transient encounters of leukocytes with the blood vessel wall. Following local damage or infection, these encounters intensify, which leads to leukocyte adhesion and subsequent migration across the endothelium, the inner lining of all blood vessels. This process is called transendothelial migration (TEM). TEM is not only the hallmark of inflammation, inflammatory disorders and cancer cell metastasis but is also required for the homing of blood stem cells to the bone marrow and lymphocyte infiltration into tumor tissue (Reymond *et al.*, 2013; van Buul *et al.*, 2002; Vestweber, 2015). TEM is a close collaboration between leukocytes on one hand and the endothelium on the other hand. For example, limiting vascular leakage not only during TEM but also when the leukocyte has crossed the endothelium is essential for keeping vascular homeostasis in check (Martinelli *et al.*, 2013; McDonald, 1994; McDonald *et al.*, 1999; Schimmel *et al.*, 2016).

The current paradigm of TEM is a refined version of the multistep model that was first proposed by Butcher and Springer (Butcher, 1991; Springer, 1994) and comprises the following successive steps: leukocyte rolling, arrest, crawling, firm adhesion, and diapedesis. The latter step occurs either through the endothelial junctions (paracellular route) (Kroon *et al.*, 2014; Schulte *et al.*, 2011) or through the endothelial cell body (transcellular route) (Carman *et al.*, 2007; Carman and Springer, 2004; Feng *et al.*, 1998).

In this chapter, we will concentrate on the role of the actin cytoskeleton of the endothelium and how this structure is of crucial importance for efficiently allowing leukocytes to cross. In particular, the crucial mediators of the actin cytoskeleton, the small GTPases, are highlighted in the process of TEM. Based on the multistep TEM model, we have classified our review in three main parts: (i) the rolling and tethering step of the leukocytes over the endothelium; (ii) the crawling and firm adhesion step; and (iii) the final diapedesis step, where the leukocytes cross the barrier (Fig. 1).

10.2. Rolling and tethering

In response to inflammatory stimulus, endothelial cells upregulate adhesion molecules to capture activated leukocytes such as neutrophils, monocytes and T-lymphocytes, and direct them to the site of inflammation (Fig. 1). The first step in this process is the rolling and tethering, accomplished through weak binding of sialyl Lewis^X-like structures on the leukocyte or endothelial surface to their selectin ligand on the opposite surface. Such an example is the mucin-like P-selectin glycoprotein ligand 1 (PSGL-1; CD162), which is constitutively expressed on leukocytes of innate immunity, and has induced expression on T-lymphocytes

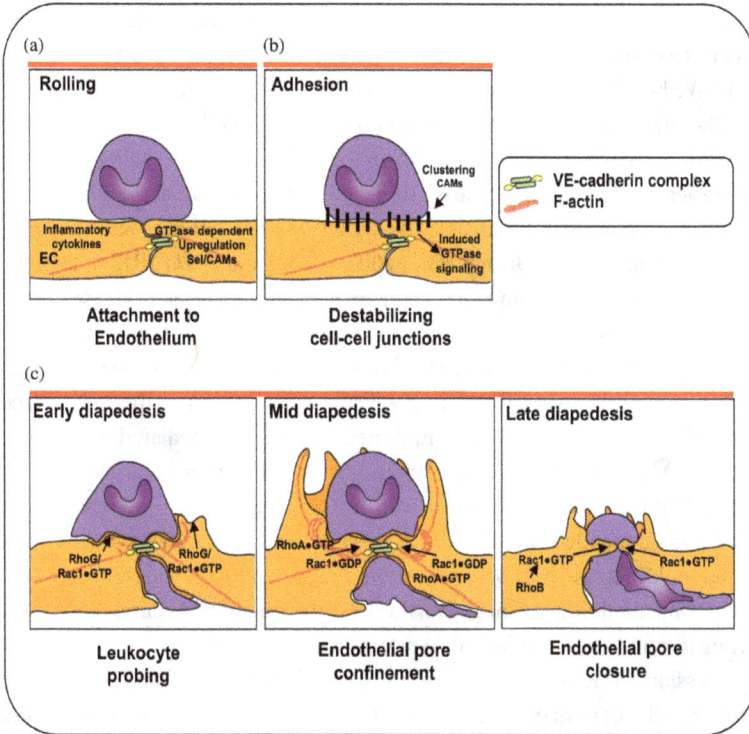

Figure 1. The leukocyte transmigration cascade — three distinct parts ((a), (b), and (c)) to cross the endothelial monolayer. (a) involves the rolling of the leukocyte over the inflamed endothelium. This locally activates the leukocytes and transits to (b), the adhesion and crawling part, where cell adhesion molecules (CAMs) on the endothelium control local stiffness, giving the leukocyte a specific surface to crawl on. When encountering a local "hot spot", the leukocyte starts its final descent: diapedesis. (c) is divided into three consecutive steps: the early-diapedesis step, where the leukocyte starts penetrating and probing the endothelial monolayer; the mid-diapedesis step, where RhoA is activated and the endothelium actively starts to prevent leakage by inducing a confined pore; and late-diapedesis step, where the endothelium needs to rapidly close the gap that is left behind by the migrating leukocyte. Rac1·GDP: inactive Rac1; Rac1·GTP: active Rac1.

(Abadier and Ley, 2017; Frenette *et al.*, 2000). It is also expressed on the surface of endothelial cells (da Costa Martins *et al.*, 2007) and has the potential to bind all members of the selectin family, including endothelial selectin (E-selectin; CD62E) and leukocyte selectin (L-selectin; CD62L), but binds with highest affinity to platelet selectin (P-selectin; CD62P) (Tinoco *et al.*, 2017).

Despite the name, P-selectin is not exclusively expressed on platelets, but like all the other selectins, it was named after where it was first discovered. P-selectin is stored in Weibel–Palade bodies (WPbs) of endothelial cells and is rapidly mobilized to the cell surface in response to acute stimuli such as histamine or thrombin (Rondaij *et al.*, 2006). These stimuli have been shown to implicate Trio in their signaling pathway (Timmerman *et al.*, 2015). Trio is a Rho GEF with two distinct GTPase-activating domains, GEF1 and GEF2 domains; the GEF1 domain activates Rac1/RhoG; and the GEF2 domain, RhoA (Blangy *et al.*, 2000; van Rijssel *et al.*, 2012a). During histamine stimulation, RhoA, and the downstream RHO associated kinases (ROCK), have been shown to become highly activated, and determine endothelial-barrier permeability (Mikelis *et al.*, 2015; van Nieuw Amerongen *et al.*, 1998). No direct evidence for involvement of RhoA in P-selectin mobilization to the cell surface has been elucidated, but it has been implicated in clathrin-mediated internalization (Setiadi and McEver, 2003; Setiadi, 2007). Rac1 on the other hand has been correlated with WPb release, a mechanism which has been shown to be reactive oxygen species (ROS) dependent (Vischer *et al.*, 1995; Yang *et al.*, 2004). Rac1 activation correlated with increase in ROS production due to NADPH oxidase activation, and overexpression of dominant-negative Rac1 or treatment with antioxidants ablated WPb translocation to the endothelial surface (Yang *et al.*, 2004).

Expression of E-selectin on the endothelial surface is induced by bacterial endotoxins and inflammatory mediators such as tumor necrosis factor-α (TNFα) and interleukin-1 (IL-1) (Schnoor, 2015). Induction leads to gene transcription and *de novo* synthesis of E-selectin, with maximum expression within 4–6 hours. The mechanism of E-selectin upregulation involves activation of necrosis factor κB (NF-κB) and c-jun terminal kinase (JNK)/p38 mitogen-activated protein kinase (p38MAPK) pathways (Read *et al.*, 1996; Schindler and Baichwal, 1994). Expression of dominant-negative RhoA, RhoB, and Rac1 has been shown to inhibit E-selectin expression on the endothelial surface (Chen *et al.*, 2003; Nubel *et al.*, 2004), all of which have been reported as activators of NF-kB (Montaner *et al.*, 1998; Perona *et al.*, 1997; Rodriguez *et al.*, 2007). RhoA has furthermore been reported as a regulator of JNK via activation of ROCK (Marinissen *et al.*, 2004), and earlier work suggested RhoA, B, and C all have this ability to activate JNK, at least in epithelial cell lines (Teramoto *et al.*, 1996). In line with the RhoA/Rac1 activating abilities of Trio, silencing of Trio resulted in decreased expression of E-selectin in TNFα-stimulated endothelial cells (van Rijssel *et al.*, 2013). Furthermore, constitutively activating Rac1 and small GTPase Cdc42 has been shown to augment E-selectin promoter activity in endothelial cells stimulated with TNFα (Min and Pober, 1997).

Clustering of P- and E-selectin around adhering leukocytes triggers important downstream effects such as changes in cell morphology, F-actin distribution, and cytosolic free calcium (Kaplanski *et al.*, 1994; Lorenzon *et al.*, 1998). Inhibiting RhoA using C3 transferase inhibits clustering of E-selectin, while the effects on P-selectin remains unclear (Wojciak-Stothard *et al.*, 1999). Interestingly, RhoA and ROCK have been demonstrated to be crucial during histamine and TNFα-stimulated endothelial cell activation (Heemskerk *et al.*, 2016b; Mikelis *et al.*, 2015), yet they seem to have opposing roles in these two conditions. Whereas RhoA inhibition preserves vascular integrity in histamine-induced endothelial activation, the opposite seems to be true when leukocytes cross TNFα-stimulated endothelial cells. In the latter case, RhoA/ROCK signaling prevents vascular leakage by inducing a local F-actin-rich ring that serves as a contractile ring. Whether this has any relation to P- and E-selectin expression and clustering remains unanswered. Nevertheless, when under normal inflammatory conditions, leukocytes are properly activated through the rolling step; they move to the second step of the multistep TEM process: firm adhesion to and crawling over the endothelium.

10.3. Firm adhesion and crawling

By bridging the transition between the initial contact and the actual diapedesis of leukocytes, the firm adhesion and crawling phase is critical for the decision if and where a leukocyte will breach the endothelial barrier (Fig. 1). The endothelium plays an important role herein by providing an adhesive and rigid surface and by guiding leukocytes to sites permissive for TEM. This is mediated by adhesion receptors, upregulated on the endothelium under inflammatory conditions, and by the F-actin cytoskeleton, which controls cellular stiffness (Tseng *et al.*, 2005). In the last two decades, it has become increasingly clear that there is intensive crosstalk between endothelial cell adhesion receptors and the cortical F-actin cytoskeleton: (i) actin-binding proteins (ABPs) connect adhesion receptors to the actin cytoskeleton, and (ii) the signaling downstream of adhesion receptors controls the dynamics of the F-actin cytoskeleton. Rho GTPases have emerged as important players in these processes by controlling F-actin cytoskeletal dynamics and therefore endothelial surface stiffness.

10.3.1. *Firm adhesion*

Chemokine-induced activation of leukocyte integrins LFA-1 ($\alpha_L\beta_2$) and VLA-4 ($\alpha_V\beta_1$) allow these adhesion receptors to interact with their endothelial counterligands intracellular cell adhesion molecule-1 (ICAM-1) and vascular cell adhesion molecule-1 (VCAM-1), respectively, resulting in firm adhesion of leukocytes to the endothelium. ICAM-1 and VCAM-1 are both members of the immunoglobulin family of adhesion receptors and become strongly upregulated by the endothelium under inflammatory conditions. On the luminal surface of inflamed endothelial cells, ICAM-1 and VCAM-1 reside together in preformed membrane nanodomains (Barreiro *et al.*, 2005, 2008). Upon binding by leukocyte integrins, these nanodomains coalesce into higher-order clusters, further supporting leukocyte firm adhesion (Barreiro *et al.*, 2005; van Buul *et al.*, 2010b). Clustering of both ICAM-1 and VCAM-1 seems to be the main mechanism for inducing signaling into the endothelial cells and was shown to result in activation of the small GTPases RhoA, Rac1, and RhoG (Adamson *et al.*, 1999; Cook-Mills *et al.*, 2004; Etienne *et al.*, 1998; Heemskerk *et al.*, 2016b; Lessey-Morillon *et al.*, 2014; Persidsky *et al.*, 2006; Ramirez *et al.*, 2008; Schnoor *et al.*, 2011; Thompson *et al.*, 2002; van Buul *et al.*, 2007a; van Rijssel *et al.*, 2012b; van Wetering *et al.*, 2003; Vockel and Vestweber, 2013). Additionally, clustering induces interaction with a number of membrane-organizing proteins, e.g., caveolin-1, annexin A2, and actin-binding proteins (ABPs), e.g., filamin, α-actinin, cortactin, which link ICAM-1 and VCAM-1 to the F-actin cytoskeleton or to specific membrane domains (Carpen *et al.*, 1992; Celli *et al.*, 2006; Heemskerk *et al.*, 2016a, b; Kanters *et al.*, 2008; Millan *et al.*, 2006; Tilghman and Hoover, 2002; van Buul *et al.*, 2010a; van Rijssel *et al.*, 2012b).

Multiple studies have shown that clustering of ICAM-1 results in the translocation of ICAM-1 into lipid raft-related/cholesterol-enriched membrane domains, such as caveolin-1-positive caveolae (reviewed by Heemskerk *et al.*, (2014)). Treatment with a cholesterol-depleting agent (methyl-β-cyclodextrin) reduced the adhesive function of ICAM-1 (van Buul *et al.*, 2010a) and coclustering of VCAM-1 with ICAM-1 (van Buul *et al.*, 2010b). These data demonstrate the importance of these domains in ICAM-1 function. Interestingly, silencing of the membrane-binding protein annexin A2 expression prevents clustering-induced translocation of ICAM-1 into caveolae. As a result, this leads to an increase in ICAM-1-dependent adhesion of neutrophils (Heemskerk *et al.*, 2016a). Membrane distribution of ICAM-1 is crucial for the ICAM-1 adhesive function by balancing its adhesive capacity and may therefore also function to limit excessive leukocyte attachment. Interestingly, caveolin-1 was demonstrated to have a negative regulatory effect on the GTPase Rac1 (Cerezo *et al.*, 2009; Nethe *et al.*, 2010, 2012), suggesting that the distribution of ICAM-1 into specific membrane compartments may also regulate and fine-tune small Rho GTPase activation in endothelial cells.

Since the intracellular domain of ICAM-1 is only 28 amino acids long, interacting ABPs are in particular important for connecting and anchoring ICAM-1 to the actin cytoskeleton. However, at least eight different ABPs (i.e., filamin A/B/C, α-actinin1/4, cortactin, ezrin, moesin) have thus far been reported to bind either directly or indirectly to this intracellular domain, suggesting that these interactions must be spatiotemporally regulated (Barreiro et al., 2002; Carpen et al., 1992; Celli et al., 2006; Heiska et al., 1998; Kanters et al., 2008; Oh et al., 2007; Romero et al., 2002; Schaefer et al., 2014; Schnoor et al., 2011; Tilghman and Hoover, 2002; van Buul et al., 2010a; van Rijssel et al., 2012b). Schaefer and co-workers corroborated this by showing that filamin B, α-actinin-4, and cortactin indeed form independent distinct molecular complexes with ICAM-1 (Schaefer et al., 2014). How these interactions are exactly regulated is still largely unclear. The F-actin destabilizing drug cytochalasin B was shown to stimulate cortactin binding but reduce filamin and α-actinin-4 interaction with ICAM-1. Alternatively, inhibition of F-actin ring formation by pharmacological targeting of the Rac1- and RhoG-activating GEF Trio reduced α-actinin-4 and cortactin recruitment to sites of ICAM-1 clustering, whereas filamin association was unaffected (van Rijssel et al., 2012b). Actin cytoskeletal dynamics itself thus seems to be an important regulator of ABP binding to ICAM-1. Conversely, each of these ABPs has also been shown to regulate actin dynamics in a different way. Through dimerization and their flexible hinges, filamins can crosslink F-actin into gel-like networks (Stossel et al., 2001). α-Actinins are involved in crosslinking F-actin into bundles (Courson and Rock, 2010), whereas cortactin acts together with the Arp2/3 complex to initiate local F-actin polymerization (Kirkbride et al., 2011). Surprisingly, these three ABPs were all shown to be required for efficient leukocyte adhesion (Kanters et al., 2008; Schaefer et al., 2014; Schnoor et al., 2011). By mediating the connection of ICAM-1 to these different actin networks, one of functions of these ABPs may therefore be to permit leukocyte firm adhesion on specific subcellular locations on the endothelial apical surface.

Besides linking adhesion receptors to the F-actin cytoskeleton, these ABPs were also shown to have important functions as adaptor proteins which facilitate Rho GTPase signaling. Initial reports studying signaling induced upon ICAM-1 and VCAM-1 clustering or antibody-mediated crosslinking demonstrated that ICAM-1 was a potent activator of RhoA (Adamson et al., 1999; Etienne et al., 1998; Heemskerk et al., 2016b; Lessey-Morillon et al., 2014; Persidsky et al., 2006; Ramirez et al., 2008; Thompson et al., 2002), whereas VCAM-1 signaling led to activation of Rac1 (Cook-Mills et al., 2004; van Wetering et al., 2003; Vockel and Vestweber, 2013). Prolonged ICAM-1 clustering resulted in F-actin stress fiber formation and gaps in endothelial monolayers, which were believed to facilitate leukocyte TEM. Alternatively, VCAM-1-mediated Rac1 activation was shown to be required for production of ROS, which through inhibition of tyrosine phosphatases and activation of matrix metalloproteinases (MMPs) could contribute to the TEM process. Later studies used beads coated with ICAM-1 antibody to cluster ICAM-1 more locally and found that clustering also led to the activation of RhoG and Rac1, which was mediated by the GEFs, SGEF, and Trio (Schnoor *et al.*, 2011; van Buul *et al.*, 2007b; van Rijssel *et al.*, 2012b). Both Trio and Rac1 were reported as binding partners of ABP filamin (Bellanger *et al.*, 2000). Trio-mediated RhoG and Rac1 activation appeared to be dependent on filamin (van Rijssel *et al.*, 2012b), demonstrating that filamin indeed functions as a scaffolding molecule for ICAM-1 signaling. In addition to filamin, also cortactin was shown to be required for RhoG activation (Schnoor *et al.*, 2011). It is therefore tempting to speculate that filamin may scaffold the Trio–RhoG/Rac1 signaling axis, whereas cortactin may be involved in SGEF-mediated RhoG activation.

Although these Rho GTPase signaling pathways were activated upon adhesion-induced clustering of ICAM-1 and VCAM-1, their contribution to leukocyte firm adhesion is questionable. Knockdown of Rac1 or RhoG expression or use of dominant-negative constructs did not affect leukocyte adhesion (van Buul *et al.*, 2007a; Wojciak-Stothard *et al.*, 1999). After inhibiting Rho using C3 transferase or a dominant-negative construct, Wojciak-Stothard and co-workers did observe reduced monocyte adhesion and spreading on endothelial cells (Wojciak-Stothard *et al.*, 1999). Nevertheless, using these same tools, others found no effects on neutrophil, monocyte, or T-cell adhesion to the endothelium (Adamson *et al.*, 1999; Carman and Springer, 2004; Saito *et al.*, 2002; Strey *et al.*, 2002). This is actually not too surprising considering that for most leukocytes the firm adhesion phase is not much more than a quick transition from rolling to crawling on the endothelium. The timing of Rho GTPase signaling induced upon adhesion receptor clustering may therefore coincide more with and regulate the consequent steps in the leukocyte extravasation cascade: crawling and diapedesis (Muller, 2015).

10.3.2. *Crawling*

After establishing firm adhesion, crawling to a site permissive for transmigration is the next step in the sequel. Leukocytes are known to preferably migrate in the direction of increasing stiffness, a process called durotaxis (Schaefer and Hordijk, 2015; Schimmel *et al.*, 2016). Endothelial cells were shown to have an increasing central-to-peripheral stiffness gradient, which was largely dependent on the ICAM-1-interacting ABPs α-actinin-4 and cortactin (Schaefer *et al.*, 2014) and most likely guides leukocytes to a suitable diapedesis site. Martinelli and colleagues proposed that leukocytes tend to choose the path of least resistance and finally transmigrate at sites of low endothelial stiffness (Martinelli *et al.*, 2014). Using podosome-like protrusions, leukocytes were shown to probe the endothelium for sites of local low stiffness, which would aid leukocytes in pathfinding (Carman *et al.*, 2007; Martinelli *et al.*, 2014; Shulman *et al.*, 2009). Interestingly, T-cells were also shown to use invasive protrusions to encounter chemokines presented by the endothelium (Shulman *et al.*, 2012). Therefore, these structures could function in directing migration by sensing stiffness (durotaxis) as well as a chemokine-induced migration (chemotaxis).

Alternatively, leukocytes can also affect endothelial cell stiffness by inducing signaling through mechanosensitive adhesion receptors on the endothelium. Lessey-Morillon and co-workers showed that ICAM-1 can function as a mechanotransducer, which upon increased pulling forces, promoted endothelial cell stiffness (Lessey-Morillon *et al.*, 2014). This mechanoresponse involved activation of RhoA via its GEF ARHGEF12/LARG and was dependent on actomyosin-based contractility. Depletion of LARG resulted in reduced leukocyte crawling speed, demonstrating that this signaling pathway indeed contributes to stiffness-dependent leukocyte crawling. Once the leukocyte finds a "hot spot" to transmigrate, it starts to overcome its last fortress: crossing the endothelial barrier or in other words the diapedesis step.

10.4. Diapedesis

The final and maybe most complicated step of the multistep process of leukocyte TEM is the actual diapedesis where leukocytes breach through the endothelial cell barrier in order to continue their way towards underlying tissues. Intensifying interactions between the two involved cell types during this stage is necessary to ensure tight regulation of critical processes like opening of the endothelial barrier layer and limiting vascular leakage (Fig. 1). Endothelial actin remodeling is a key event during leukocyte diapedesis as the endothelial cells need to reorganize either their cytoplasmic structure or their intercellular junctional connections in order to allow the transmigrating leukocyte to pass via the transcellular or paracellular route (Schnoor, 2015). The role of actin and its regulating GTPases during the transition from adhesive to initiating diapedesis, the opening of endothelial cell-cell junctions, the preservation of vascular permeability during leukocyte diapedesis, and finally the resealing of the endothelial barrier afterwards will be discussed below.

10.4.1. *Apical membrane structures*

The transition from leukocyte adhesion to the actual diapedesis is regulated by a process that involves endothelial lateral protrusions. This process goes with many different names that all describe the same phenomenon: the surrounding of the migrating leukocytes by an endothelial membrane protrusion. The first study that reported these structures described them as docking structures (Barreiro *et al.*, 2002, 2004), shortly followed by the so-called transmigratory cup studies (Carman *et al.*, 2003; Carman and Springer, 2004). Other terms known for these structures are apical cups (van Buul *et al.*, 2007a), dome structures (Petri *et al.*, 2008, 2011; Phillipson *et al.*, 2008), actin dynamic structures (Mooren *et al.*, 2014) or ICAM-1-rich contact areas (Schnoor *et al.*, 2011; Vestweber *et al.*, 2013). Their hypothesized function is still debated: these structures may function as adhesion platforms but may also be necessary for supporting transmigration by extending the endothelial membrane. Moreover, they may function in conjunction with the F-actin-rich contractile ring to limit vascular leakage (Heemskerk *et al.*, 2016b; Schimmel *et al.*, 2016). Additionally, the formation of these apical protrusions might in the end facilitate the transition of leukocytes from adhesion to the last stage of extravasation: diapedesis.

Taking a closer look at these actin structures, there are actually two different actin structures present: vertical microvilli-like protrusions towards the apical surface and an F-actin dense ring surrounding transmigrating cells that is localized more at the basal site of endothelial cells (Fig. 1(c)) (Heemskerk *et al.*, 2016b; Schimmel *et al.*, 2016). The microvilli-like protrusions are formed by polymerized actin filaments that requires activation by RhoG and Rac1 (Schnoor *et al.*, 2011; van Buul *et al.*, 2007a; van Rijssel *et al.*, 2012b) and contain endothelial adhesion receptors like ICAM-1 and VCAM-1 and present proinflammatory chemokines like IL-8 (Middleton *et al.*, 1997; Whittall *et al.*, 2013), thereby providing a platform for leukocytes to adhere to and are therefore considered to be part of the adhesion step.

The other actin structure present on the more basal side of the observed apical protrusions is described to be involved in limiting vascular leakage when leukocytes breach through the endothelial layer and therefore play a role in the actual diapedesis (Heemskerk *et al.*, 2016b). Additionally, recently, it became evident that leukocytes themselves can breach their way through these actin structures by pushing their nucleus forward and use their nucleus as a pushing force to "drill" their way through without harming the actual actin or microtubule structures (Barzilai *et al.*, 2017). These two mechanisms, F-actin rings and displacement of actin structures by leukocytes, seem to work in conjunction. How these structures are controlled by small GTPases and how GTPases mediate endothelial cell–cell junctions during TEM will be discussed below.

10.4.2. RhoA-mediated limiting of vascular permeability during leukocyte TEM

Endothelial RhoA has been implicated in the leukocyte adhesion step. This assumption is mainly based on ICAM-1 crosslinking studies where antibodies or anti-ICAM-1-coated beads were used. At the time, these techniques were required to measure RhoA activity with classical biochemical pull-down assays. Unfortunately, the downside of this approach is that you lack any information on spatial and temporal activation. This spatiotemporal information is in particular important for such a delicate and local process as leukocyte transendothelial migration (Goswami and Vestweber, 2016; Muller, 2015; Vestweber, 2015).

Making use of an optimized and validated FRET-based DORA biosensor to investigate activation of endothelial RhoA during leukocyte transendothelial migration in both time and subcellular location, we show specific RhoA activity around the site of transmigration right after the leukocytes breach through the endothelial cell layer (mid-diapedesis step) until it is almost completely underneath the endothelial cells (late-diapedesis step; Fig. 2) (Heemskerk et al., 2016b). Interestingly, we did not detect any activation of RhoA during the adhesion and crawling part or early-diapedesis step. This was surprising, since based on previous studies, it was suggested that during the early-diapedesis step, ICAM-1-induced RhoA activation would lead to opening of endothelial cell–cell junctions, basically following the same signaling pathways that were observed when endothelial cells were treated with thrombin or histamine (van Nieuw Amerongen et al., 2000). However, thrombin-induced RhoA and leukocyte-mediated RhoA activation may not result in the same functional outcome, i.e., the loss of cell–cell junctions. Whereas thrombin induces tension in endothelial cells, as is evidenced by the presence of strong and prominent F-actin stress fibers, leukocyte adhesion does not induce such prominent structures. This indicates that the data obtained with artificial ICAM-1 clustering using antibodies and antibody-coated beads may have been overinterpreted. Using such approaches, the signals induced into the endothelium may be overwhelming whereas a single leukocyte would crosslink much less ICAM-1 molecules and thereby not stimulate RhoA to the extent that was believed to occur previously. We wish to stress that local RhoA activation downstream from ICAM-1 clustering may still occur. However, we were not able to detect this with our FRET-based biosensor. Additional proof that Rho signaling is not required for opening of cell–cell junctions comes from studies that show that inhibiting tension in endothelial cells by pretreatment with Rho/Rho kinase inhibitors or depleting the cells for RhoA did not prevent leukocytes from crossing (Barzilai *et al.*, 2017; Heemskerk *et al.*, 2016b). Instead, blocking RhoA signaling in endothelial cells did result in an increase in vascular permeability when leukocytes crossed the endothelium, both *in vitro* and *in vivo* (Heemskerk *et al.*, 2016b).

Based on these recent findings with the spatiotemporal activation of RhoA during leukocyte transendothelial migration, we concluded that endothelial RhoA induced local F-actin-rich rings around a migrating leukocyte that contribute to local endothelial pore confinement and helps to maintain the endothelial barrier integrity during leukocyte diapedesis (Schimmel *et al.*, 2016). Initiation of this RhoA activity demands precise timing and localization which involves ICAM-1 and the RhoGEFs Ect2 and LARG (Fig. 2). Not only clustering of ICAM-1 leads to recruitment of Ect2 and LARG to the intracellular tail of ICAM-1 (Heemskerk *et al.*, 2016b), also the mechanical forces exerted on endothelial ICAM-1 by pulling and pushing of passing leukocytes result in LARG recruitment and finally activation of RhoA (Lessey-Morillon *et al.*, 2014). Downstream of RhoA, ROCK and myosin light chain kinase (MLCK) result in myosin-II phosphorylation and subsequent contraction of two opposing actin filaments (Vicente-Manzanares *et al.*, 2009). Because of these molecular mechanisms, leukocyte-induced F-actin rings serve as an electric strap that serves to minimize the size of the formed actin pore and as a consequence limit vascular permeability (Fig. 2).

Involvement of other GTPases in limiting vascular permeability during leukocyte diapedesis remains to be elucidated. *In vivo* inhibition of RhoA, RhoB, and RhoC with C3 transferase shows similar effects as RhoA depletion in cultured endothelial cells (Heemskerk *et al.*, 2016b), suggesting that RhoB and RhoC might not be necessary for the formation of the F-actin-rich pore that surrounds passing leukocytes.

10.5. Rac1 driven closure of endothelial gaps after leukocyte TEM

Even though the RhoA-driven F-actin ring tightly surrounds transmigrating leukocytes, there is evidence that pore closure does not happen solely via contraction of this ring, the so-called "purse string" mechanism. Dynamic actin bursts in lamellipodia-like structures are observed to close mechanical induced microwounds in an asymmetrical manner in endothelial cells (Martinelli *et al.*, 2013). Also, spontaneous loss of cell–cell contacts is rapidly followed by ventral lamellipodial

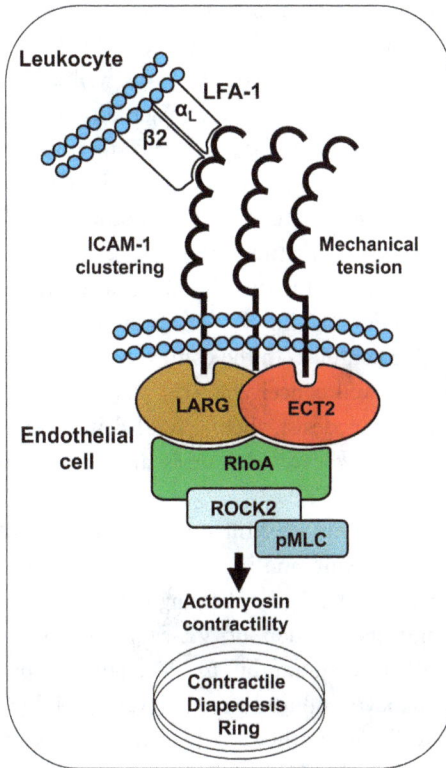

Figure 2. Signaling cascade downstream from leukocyte migration. ICAM-1 clustering through leukocyte integrin beta2 recruit the RhoGEFs LARG and Ect2, resulting in local activation of RhoA. RhoA downstream promotes ROCK2 activation and myosin-light-chain phosphorylation (pMLC) to induce actomyosin-based contraction of a F-actin-rich ring around the penetrating leukocyte serving as an elastic strap to limit vascular leakage upon leukocyte transmigration.

activity (personal observation, JDvB). These ventral lamellipodia show F-actin-rich structures underneath the endothelium at a focal plane different from the one at which RhoA-induced F-actin ring was observed. Dominant-negative Rac1 over-expression and Rac1 inhibition with NSC23766 decreased or arrested ventral lamellipodia formation and increased gap closure time, in contrast to RhoA deple-tion or inhibition of RhoA effector ROCK with the inhibitor Y27632 that showed no effect on ventral lamellipodial activity. Together with enrichment of Rac1 effec-tors cortactin, IQGAP, and p47phox in the ventral lamellipodia, Martinelli and coworkers stated that the formation of Rac1-driven ventral lamellipodia rapidly restores the endothelial barrier upon injury (Martinelli *et al.*, 2013). Independent from the leukocyte transmigration route, para- as well as transcellular gaps are closed by ventral lamellipodia. However, a major difference between para- and transcellular gap closures is the need for junction restoration in the case of paracellular transmigration. VE-cadherin dynamics, the major component of endothelial cell–cell junctions, and thereby restoration of endothelial cell junctions are regulated by the Rac1-regulated actin nucleator ARP2/3 (Abu Taha *et al.*, 2014; Mooren *et al.*, 2014). Both ventral lamellipodia that close endothelial gaps and maintain endothelial junction stability by VE-cadherin regulation depend on the actin controlling ARP2/3 complex. In the context of vascular permeability during leukocyte diapedesis, it is well conceivable that RhoA-driven F-actin pores limit leakage during leukocyte diapedesis, whereas Rac1 drives ventral lamellipodia that close the gap right after leukocyte diapedesis and restore endothelial junction stability in case of paracellular TEM.

Recently, elegant data from the Millan lab showed that RhoB is required to bring Rac1 to the plasma membrane in order to close the gap (Marcos-Ramiro *et al.*, 2016). Our data add to that by implicating the RhoGEF Trio being required for the maturation and stabilization of VE-cadherin-based cell–cell junctions (Timmerman *et al.*, 2015). It is tempting to speculate that the same machinery is in place to close the leukocyte-induced gaps. However, solid proof to confirm this hypothesis is not yet available and thus would require future research.

FRET-based biosensor data revealed that RhoA activity may take place at the previously described apical membrane structures, the exact same location where activation of other GTPases RhoG and Rac1 may take place. How come all these GTPases are regulated differently and yet localize at the same location? Different spatial activation patterns for Rac1, Rac2, and Cdc42 were shown with FRET-based biosensors in processes such as phagocytosis (Hoppe and Swanson, 2004) and micropinocytosis (deBakker *et al.*, 2004; Ellerbroek *et al.*, 2004). These processes use similar molecular players like endothelial cells for the formation of the apical membrane structures. Unpublished data from our lab using a FRET-based DORA Rac1 biosensor indicates local Rac1 activation during the final part of diapedesis when the leukocyte is almost completely through the endothelial layer (late-diapedesis step). And although many efforts have been made during the last couple of years, we still know relatively little on the true contribution of the small GTPases in the process of leukocyte transmigration. Most studies have focused on RhoA and Rac1. However, the role for Cdc42 and other family members such as Rnd and RhoBTB and others is completely unknown (Wennerberg and Der, 2004).

10.6. Summary

When all these findings are taken together, an emerging picture arises that implicates the involvement of different GTPases at different time points and different locations in the endothelial cell during leukocyte transendothelial migration. Intriguingly, one GTPase can have more than one functional outcome. These realizations put an extra level of complexity on the regulation of transendothelial migration. Therefore, future research focus may shift toward the more upstream regulators of the small GTPases: the so-called Rho guanine nucleotide exchange factors (GEFs). These proteins determine the "go" or "no-go" for GTPase activity (Rossman *et al.*, 2005) and most likely determine the local functional outcome of the GTPases during leukocyte transendothelial migration, because of their specific localization. In the end, we have only started to understand the basics of how small GTPases regulate the complex process of leukocyte transmigration. Future evidence will slowly but surely unveil the complete picture.

References

Abadier, M. and Ley, K. (2017). P-selectin glycoprotein ligand-1 in T cells. Curr. Opin. Hematol. *24*, 265–273.

Abu Taha, A., Taha, M., Seebach, J., and Schnittler, H.J. (2014). ARP2/3-mediated junction-associated lamellipodia control VE-cadherin-based cell junction dynamics and maintain monolayer integrity. Mol. Biol. Cell *25*, 245–256.

Adamson, P., Etienne, S., Couraud, P.O., Calder, V., and Greenwood, J. (1999). Lymphocyte migration through brain endothelial cell monolayers involves signaling through endothelial ICAM-1 via a rho-dependent pathway. J. Immunol. *162*, 2964–2973.

Barreiro, O., Vicente-Manzanares, M., Urzainqui, A., Yanez-Mo, M., and Sanchez-Madrid, F. (2004). Interactive protrusive structures during leukocyte adhesion and transendothelial migration. Front. Biosci. *9*, 1849–1863.

Barreiro, O., Yanez-Mo, M., Sala-Valdes, M., Gutierrez-Lopez, M.D., Ovalle, S., Higginbottom, A., Monk, P.N., Cabanas, C., and Sanchez-Madrid, F. (2005). Endothelial tetraspanin microdomains regulate leukocyte firm adhesion during extravasation. Blood *105*, 2852–2861.

Barreiro, O., Yanez-Mo, M., Serrador, J.M., Montoya, M.C., Vicente-Manzanares, M., Tejedor, R., Furthmayr, H., and Sanchez-Madrid, F. (2002). Dynamic interaction of VCAM-1 and ICAM-1 with moesin and ezrin in a novel endothelial docking structure for adherent leukocytes. J. Cell Biol. *157*, 1233–1245.

Barreiro, O., Zamai, M., Yanez-Mo, M., Tejera, E., Lopez-Romero, P., Monk, P.N., Gratton, E., Caiolfa, V.R., and Sanchez-Madrid, F. (2008). Endothelial adhesion receptors are recruited to adherent leukocytes by inclusion in preformed tetraspanin nanoplatforms. J. Cell Biol. *183*, 527–542.

Barzilai, S., Yadav, S.K., Morrell, S., Roncato, F., Klein, E., Stoler-Barak, L., Golani, O., Feigelson, S.W., Zemel, A., Nourshargh, S., et al. (2017). Leukocytes breach endothelial barriers by insertion of nuclear lobes and disassembly of endothelial actin filaments. Cell Rep. *18*, 685–699.

Bellanger, J.M., Astier, C., Sardet, C., Ohta, Y., Stossel, T.P., and Debant, A. (2000). The Rac1- and RhoG-specific GEF domain of Trio targets filamin to remodel cytoskeletal actin. Nat. Cell Biol. *2*, 888–892.

Blangy, A., Vignal, E., Schmidt, S., Debant, A., Gauthier-Rouviere, C., and Fort, P. (2000). TrioGEF1 controls Rac- and Cdc42-dependent cell structures through the direct activation of rhoG. J. Cell Sci. *113* (Pt *4*), 729–739.

Butcher, E.C. (1991). Leukocyte-endothelial cell recognition: three (or more) steps to specificity and diversity. Cell *67*, 1033–1036.

Carman, C.V., Jun, C.D., Salas, A., and Springer, T.A. (2003). Endothelial cells proactively form microvilli-like membrane projections upon intercellular adhesion molecule 1 engagement of leukocyte LFA-1. J. Immunol. *171*, 6135–6144.

Carman, C.V., Sage, P.T., Sciuto, T.E., de la Fuente, M.A., Geha, R.S., Ochs, H.D., Dvorak, H.F., Dvorak, A.M., and Springer, T.A. (2007). Transcellular diapedesis is initiated by invasive podosomes. Immunity *26*, 784–797.

Carman, C.V. and Springer, T.A. (2004). A transmigratory cup in leukocyte diapedesis both through individual vascular endothelial cells and between them. J. Cell Biol. *167*, 377–388.

Carpen, O., Pallai, P., Staunton, D.E., and Springer, T.A. (1992). Association of intercellular adhesion molecule-1 (ICAM-1) with actin-containing cytoskeleton and alpha-actinin. J. Cell Biol. *118*, 1223–1234.

Celli, L., Ryckewaert, J.J., Delachanal, E., and Duperray, A. (2006). Evidence of a functional role for interaction between ICAM-1 and nonmuscle alpha-actinins in leukocyte diapedesis. J. Immunol. *177*, 4113–4121.

Cerezo, A., Guadamillas, M.C., Goetz, J.G., Sanchez-Perales, S., Klein, E., Assoian, R.K., and del Pozo, M.A. (2009). The absence of caveolin-1 increases proliferation and anchorage- independent growth by a Rac-dependent, Erk-independent mechanism. Mol. Cell Biol. *29*, 5046–5059.

Chen, X.L., Zhang, Q., Zhao, R., Ding, X., Tummala, P.E., and Medford, R.M. (2003). Rac1 and superoxide are required for the expression of cell adhesion molecules induced by tumor necrosis factor-α in endothelial cells. J. Pharmacol. Exp. Ther. *305*, 573–580.

Cook-Mills, J.M., Johnson, J.D., Deem, T.L., Ochi, A., Wang, L., and Zheng, Y. (2004). Calcium mobilization and Rac1 activation are required for VCAM-1 (vascular cell adhesion molecule-1) stimulation of NADPH oxidase activity. Biochem. J. *378*, 539–547.

Courson, D.S. and Rock, R.S. (2010). Actin cross-link assembly and disassembly mechanics for alpha-actinin and fascin. J. Biol. Chem. *285*, 26350–26357.

da Costa Martins, P., Garcia-Vallejo, J.J., van Thienen, J.V., Fernandez-Borja, M., van Gils, J.M., Beckers, C., Horrevoets, A.J., Hordijk, P.L., and Zwaginga, J.J. (2007).

P-selectin glycoprotein ligand-1 is expressed on endothelial cells and mediates monocyte adhesion to activated endothelium. Arterioscler. Thromb. Vasc. Biol. *27*, 1023–1029.

deBakker, C.D., Haney, L.B., Kinchen, J.M., Grimsley, C., Lu, M., Klingele, D., Hsu, P.K., Chou, B.K., Cheng, L.C., Blangy, A., *et al.* (2004). Phagocytosis of apoptotic cells is regulated by a UNC-73/TRIO-MIG-2/RhoG signaling module and armadillo repeats of CED-12/ELMO. Curr. Biol. *14*, 2208–2216.

Ellerbroek, S.M., Wennerberg, K., Arthur, W.T., Dunty, J.M., Bowman, D.R., DeMali, K.A., Der, C., and Burridge, K. (2004). SGEF, a RhoG guanine nucleotide exchange factor that stimulates macropinocytosis. Mol. Biol. Cell *15*, 3309–3319.

Etienne, S., Adamson, P., Greenwood, J., Strosberg, A.D., Cazaubon, S., and Couraud, P.O. (1998). ICAM-1 signaling pathways associated with Rho activation in microvascular brain endothelial cells. J. Immunol. *161*, 5755–5761.

Feng, D., Nagy, J.A., Pyne, K., Dvorak, H.F., and Dvorak, A.M. (1998). Neutrophils emigrate from venules by a transendothelial cell pathway in response to FMLP. J. Exp. Med. *187*, 903–915.

Frenette, P.S., Denis, C.V., Weiss, L., Jurk, K., Subbarao, S., Kehrel, B., Hartwig, J.H., Vestweber, D., and Wagner, D.D. (2000). P-selectin glycoprotein ligand 1 (PSGL-1) is expressed on platelets and can mediate platelet-endothelial interactions *in vivo*. J. Exp. Med. *191*, 1413–1422.

Goswami, D. and Vestweber, D. (2016). How leukocytes trigger opening and sealing of gaps in the endothelial barrier. F1000Research *5*, 2321. doi: 10.12688/f1000research.9185.1.

Heemskerk, N., Asimuddin, M., Oort, C., van Rijssel, J., and van Buul, J.D. (2016a). Annexin A2 limits neutrophil transendothelial migration by organizing the spatial distribution of ICAM-1. J. Immunol. *196*, 2767–2778.

Heemskerk, N., Schimmel, L., Oort, C., van Rijssel, J., Yin, T., Ma, B., van, U.J., Pitter, B., Huveneers, S., Goedhart, J., *et al.* (2016b). F-actin-rich contractile endothelial pores prevent vascular leakage during leukocyte diapedesis through local RhoA signalling. Nat. Commun. *7*, 10493.

Heemskerk, N., van Rijssel, J., and van Buul, J.D. (2014). Rho-GTPase signaling in leukocyte extravasation: an endothelial point of view. Cell Adhes. Migr. *8*, 67–75.

Heiska, L., Alfthan, K., Gronholm, M., Vilja, P., Vaheri, A., and Carpen, O. (1998). Association of ezrin with intercellular adhesion molecule-1 and -2 (ICAM-1 and ICAM-2). Regulation by phosphatidylinositol 4, 5-bisphosphate. J. Biol. Chem. *273*, 21893–21900.

Hoppe, A.D. and Swanson, J.A. (2004). Cdc42, Rac1, and Rac2 display distinct patterns of activation during phagocytosis. Mol. Biol. Cell *15*, 3509–3519.

Kanters, E., van Rijssel, J., Hensbergen, P.J., Hondius, D., Mul, F.P., Deelder, A.M., Sonnenberg, A., van Buul, J.D., and Hordijk, P.L. (2008). Filamin B mediates ICAM-1-driven leukocyte transendothelial migration. J. Biol. Chem. *283*, 31830–31839.

Kaplanski, G., Farnarier, C., Benoliel, A.M., Foa, C., Kaplanski, S., and Bongrand, P. (1994). A novel role for E- and P-selectins: shape control of endothelial cell monolayers. J. Cell Sci. *107* (*Pt 9*), 2449–2457.

Kirkbride, K.C., Sung, B.H., Sinha, S., and Weaver, A.M. (2011). Cortactin: a multifunctional regulator of cellular invasiveness. Cell Adhes. Migr. *5*, 187–198.

Kroon, J., Daniel, A.E., Hoogenboezem, M., and van Buul, J.D. (2014). Real-time imaging of endothelial cell-cell junctions during neutrophil transmigration under physiological flow. J. Vis. Exp. *90*, 51766. doi: 10.3791/51766.

Lessey-Morillon, E.C., Osborne, L.D., Monaghan-Benson, E., Guilluy, C., O'Brien, E.T., Superfine, R., and Burridge, K. (2014). The RhoA guanine nucleotide exchange factor, LARG, mediates ICAM-1-dependent mechanotransduction in endothelial cells to stimulate transendothelial migration. J. Immunol. *192*, 3390–3398.

Lorenzon, P., Vecile, E., Nardon, E., Ferrero, E., Harlan, J.M., Tedesco, F., and Dobrina, A. (1998). Endothelial cell E- and P-selectin and vascular cell adhesion molecule-1 function as signaling receptors. J. Cell Biol. *142*, 1381–1391.

Marcos-Ramiro, B., Garcia-Weber, D., Barroso, S., Feito, J., Ortega, M.C., Cernuda-Morollon, E., Reglero-Real, N., Fernandez-Martin, L., Duran, M.C., Alonso, M.A., et al. (2016). RhoB controls endothelial barrier recovery by inhibiting Rac1 trafficking to the cell border. J. Cell Biol. *213*, 385–402.

Marinissen, M.J., Chiariello, M., Tanos, T., Bernard, O., Narumiya, S., and Gutkind, J.S. (2004). The small GTP-binding protein RhoA regulates c-jun by a ROCK-JNK signaling axis. Mol. Cell *14*, 29–41.

Martinelli, R., Kamei, M., Sage, P.T., Massol, R., Varghese, L., Sciuto, T., Toporsian, M., Dvorak, A.M., Kirchhausen, T., Springer, T.A., et al. (2013). Release of cellular tension signals self-restorative ventral lamellipodia to heal barrier micro-wounds. J. Cell Biol. *201*, 449–465.

Martinelli, R., Zeiger, A.S., Whitfield, M., Scuito, T.E., Dvorak, A., Van Vliet, K.J., Greenwood, J., and Carman, C.V. (2014). Probing the biomechanical contribution of the endothelium to lymphocyte migration: diapedesis by the path of least resistance. J. Cell Sci. *127*, 3720–3734.

McDonald, D.M. (1994). Endothelial gaps and permeability of venules in rat tracheas exposed to inflammatory stimuli. Am. J. Physiol. *266*, L61–L83.

McDonald, D.M., Thurston, G., and Baluk, P. (1999). Endothelial gaps as sites for plasma leakage in inflammation. Microcirculation *6*, 7–22.

Middleton, J., Neil, S., Wintle, J., Clark-Lewis, I., Moore, H., Lam, C., Auer, M., Hub, E., and Rot, A. (1997). Transcytosis and surface presentation of IL-8 by venular endothelial cells. Cell *91*, 385–395.

Mikelis, C.M., Simaan, M., Ando, K., Fukuhara, S., Sakurai, A., Amornphimoltham, P., Masedunskas, A., Weigert, R., Chavakis, T., Adams, R.H., et al. (2015). RhoA and ROCK mediate histamine-induced vascular leakage and anaphylactic shock. Nat. Commun. *6*, 6725.

Millan, J., Hewlett, L., Glyn, M., Toomre, D., Clark, P., and Ridley, A.J. (2006). Lymphocyte transcellular migration occurs through recruitment of endothelial ICAM-1 to caveola- and F-actin-rich domains. Nat. Cell Biol. *8*, 113–123.

Min, W. and Pober, J.S. (1997). TNF initiates E-selectin transcription in human endothelial cells through parallel TRAF-NF-kappa B and TRAF-RAC/CDC42-JNK-c-Jun/ATF2 pathways. J. Immunol. *159*, 3508–3518.

Montaner, S., Perona, R., Saniger, L., and Lacal, J.C. (1998). Multiple signalling pathways lead to the activation of the nuclear factor κB by the Rho family of GTPases. J. Biol. Chem. *273*, 12779–12785.

Mooren, O.L., Li, J., Nawas, J., and Cooper, J.A. (2014). Endothelial cells use dynamic actin to facilitate lymphocyte transendothelial migration and maintain the monolayer barrier. Mol. Biol. Cell *25*, 4115–4129.

Muller, W.A. (2015). Localized signals that regulate transendothelial migration. Curr. Opin. Immunol. *38*, 24–29.

Nethe, M., Anthony, E.C., Fernandez-Borja, M., Dee, R., Geerts, D., Hensbergen, P.J., Deelder, A.M., Schmidt, G., and Hordijk, P.L. (2010). Focal-adhesion targeting links caveolin-1 to a Rac1-degradation pathway. J. Cell Sci. *123*, 1948–1958.

Nethe, M., de Kreuk, B.J., Tauriello, D.V., Anthony, E.C., Snoek, B., Stumpel, T., Salinas, P.C., Maurice, M.M., Geerts, D., Deelder, A.M., et al. (2012). Rac1 acts in conjunction with Nedd4 and dishevelled-1 to promote maturation of cell–cell contacts. J. Cell Sci. *125*, 3430–3442.

Nubel, T., Dippold, W., Kleinert, H., Kaina, B., and Fritz, G. (2004). Lovastatin inhibits Rho-regulated expression of E-selectin by TNFalpha and attenuates tumor cell adhesion. FASEB J. *18*, 140–142.

Oh, H.M., Lee, S., Na, B.R., Wee, H., Kim, S.H., Choi, S.C., Lee, K.M., and Jun, C.D. (2007). RKIKK motif in the intracellular domain is critical for spatial and dynamic organization of ICAM-1: functional implication for the leukocyte adhesion and trans-migration. Mol. Biol. Cell *18*, 2322–2335.

Perona, R., Montaner, S., Saniger, L., Sanchez-Perez, I., Bravo, R., and Lacal, J.C. (1997). Activation of the nuclear factor-kappaB by Rho, CDC42, and Rac-1 proteins. Genes Dev. *11*, 463–475.

Persidsky, Y., Heilman, D., Haorah, J., Zelivyanskaya, M., Persidsky, R., Weber, G.A., Shimokawa, H., Kaibuchi, K., and Ikezu, T. (2006). Rho-mediated regulation of tight junctions during monocyte migration across the blood-brain barrier in HIV-1 encephalitis (HIVE). Blood *107*, 4770–4780.

Petri, B., Phillipson, M., and Kubes, P. (2008). The physiology of leukocyte recruitment: an *in vivo* perspective. J. Immunol. *180*, 6439–6446.

Petri, B.r., Kaur, J., Long, E.M., Li, H., Parsons, S.A., Butz, S., Phillipson, M., Vestweber, D., Patel, K.D., Robbins, S.M., et al. (2011). Endothelial LSP1 is involved in endothelial dome formation, minimizing vascular permeability changes during neutrophil transmigration *in vivo*. Blood *117*, 942–952.

Phillipson, M., Kaur, J., Colarusso, P., Ballantyne, C.M., and Kubes, P. (2008). Endothelial domes encapsulate adherent neutrophils and minimize increases in vascular permeability in paracellular and transcellular emigration. PLoS ONE *3*, e1649.

Pober, J.S. and Sessa, W.C. (2007). Evolving functions of endothelial cells in inflammation. Nat. Rev. Immunol. *7*, 803–815.

Ramirez, S.H., Heilman, D., Morsey, B., Potula, R., Haorah, J., and Persidsky, Y. (2008). Activation of peroxisome proliferator-activated receptor gamma (PPARgamma) suppresses Rho GTPases in human brain microvascular endothelial cells and inhibits adhesion and transendothelial migration of HIV-1 infected monocytes. J. Immunol. *180*, 1854–1865.

Read, M.A., Neish, A.S., Gerritsen, M.E., and Collins, T. (1996). Postinduction transcriptional repression of E-selectin and vascular cell adhesion molecule-1. J. Immunol. *157*, 3472–3479.

Reymond, N., d'Agua, B.B., and Ridley, A.J. (2013). Crossing the endothelial barrier during metastasis. Nat. Rev. Cancer *13*, 858–870.

Rodriguez, R., Campa, V.M., Riera, J., Carcedo, M.T., Ucker, D.S., Ramos, S., and Lazo, P.S. (2007). TNF triggers mitogenic signals in NIH 3T3 cells but induces apoptosis when the cell cycle is blocked. Eur. Cytokine Netw. *18*, 172–180.

Romero, I.A., Amos, C.L., Greenwood, J., and Adamson, P. (2002). Ezrin and moesin co-localise with ICAM-1 in brain endothelial cells but are not directly associated. Mol. Brain Res. *105*, 47–59.

Rondaij, M.G., Bierings, R., Kragt, A., van Mourik, J.A., and Voorberg, J. (2006). Dynamics and plasticity of Weibel-Palade bodies in endothelial cells. Arterioscler. Thromb. Vasc. Biol. *26*, 1002–1007.

Rossman, K.L., Der, C.J., and Sondek, J. (2005). GEF means go: turning on RHO GTPases with guanine nucleotide-exchange factors. Nat. Rev. Mol. Cell Biol. *6*, 167–180.

Saito, H., Minamiya, Y., Saito, S., and Ogawa, J.I. (2002). Endothelial Rho and Rho kinase regulate neutrophil migration via endothelial myosin light chain phosphorylation. J. Leukoc. Biol. *72*, 829–836.

Schaefer, A. and Hordijk, P.L. (2015). Cell-stiffness-induced mechanosignaling — a key driver of leukocyte transendothelial migration. J. Cell Sci. *128*, 2221–2230.

Schaefer, A., te Riet, J., Ritz, K., Hoogenboezem, M., Anthony, E.C., Mul, F.P.J., de Vries, C.J., Daemen, M.J., Figdor, C.G., van Buul, J.D., et al. (2014). Actin-binding proteins differentially regulate endothelial cell stiffness, ICAM-1 function and neutrophil transmigration. J. Cell Sci. *127*, 4470–4482.

Schimmel, L., Heemskerk, N., and van Buul, J.D. (2016). Leukocyte transendothelial migration: a local affair. Small GTPases *8*, 1–15.

Schindler, U. and Baichwal, V.R. (1994). Three NF-kappa B binding sites in the human E-selectin gene required for maximal tumor necrosis factor alpha-induced expression. Mol. Cell Biol. *14*, 5820–5831.

Schnoor, M. (2015). Endothelial actin-binding proteins and actin dynamics in leukocyte transendothelial migration. J. Immunol. *194*, 3535–3541.

Schnoor, M., Lai, F.P.L., Zarbock, A., Kläver, R., Polaschegg, C., Schulte, D., Weich, H.A., Oelkers, J.M., Rottner, K., and Vestweber, D. (2011). Cortactin deficiency is associated with reduced neutrophil recruitment but increased vascular permeability *in vivo*. J. Exp. Med. *208*, 1721–1735.

Schulte, D., Kuppers, V., Dartsch, N., Broermann, A., Li, H., Zarbock, A., Kamenyeva, O., Kiefer, F., Khandoga, A., Massberg, S., *et al.* (2011). Stabilizing the VE-cadherin-catenin complex blocks leukocyte extravasation and vascular permeability. EMBO J. *30*, 4157–4170.

Setiadi, H. and McEver, R.P. (2003). Signal-dependent distribution of cell surface P-selectin in clathrin-coated pits affects leukocyte rolling under flow. J. Cell Biol. *163*, 1385–1395.

Setiadi, H.M. (2007). Clustering endothelial E-selectin in clathrin-coated pits and lipid rafts enhances leukocyte adhesion under flow. Blood *111*, 1989–1998.

Shulman, Z., Cohen, S.J., Roediger, B., Kalchenko, V., Jain, R., Grabovsky, V., Klein, E., Shinder, V., Stoler-Barak, L., Feigelson, S.W., *et al.* (2012). Transendothelial migration of lymphocytes mediated by intraendothelial vesicle stores rather than by extracellular chemokine depots. Nat. Immunol. *13*, 67–76.

Shulman, Z., Shinder, V., Klein, E., Grabovsky, V., Yeger, O., Geron, E., Montresor, A., Bolomini-Vittori, M., Feigelson, S.W., Kirchhausen, T., *et al.* (2009). Lymphocyte crawling and transendothelial migration require chemokine triggering of high-affinity LFA-1 integrin. Immunity *30*, 384–396.

Springer, T.A. (1994). Traffic signals for lymphocyte recirculation and leukocyte emigration: the multistep paradigm. Cell *76*, 301–314.

Stossel, T.P., Condeelis, J., Cooley, L., Hartwig, J.H., Noegel, A., Schleicher, M., and Shapiro, S.S. (2001). Filamins as integrators of cell mechanics and signalling. Nat. Rev. Mol. Cell Biol. *2*, 138–145.

Strey, A., Janning, A., Barth, H., and Gerke, V. (2002). Endothelial Rho signaling is required for monocyte transendothelial migration. FEBS Lett. *517*, 261–266.

Teramoto, H., Crespo, P., Coso, O.A., Igishi, T., Xu, N., and Gutkind, J.S. (1996). The small GTP-binding protein rho activates c-Jun N-terminal kinases/stress-activated protein kinases in human kidney 293T cells. Evidence for a Pak-independent signaling pathway. J. Biol. Chem. *271*, 25731–25734.

Thompson, P.W., Randi, A.M., and Ridley, A.J. (2002). Intercellular adhesion molecule (ICAM)-1, but not ICAM-2, activates RhoA and stimulates c-fos and rhoA transcription in endothelial cells. J. Immunol. *169*, 1007–1013.

Tilghman, R.W. and Hoover, R.L. (2002). E-selectin and ICAM-1 are incorporated into detergent-insoluble membrane domains following clustering in endothelial cells. FEBS Lett. *525*, 83–87.

Timmerman, I., Heemskerk, N., Kroon, J., Schaefer, A., van Rijssel, J., Hoogenboezem, M., van Unen, J., Goedhart, J., Gadella, T.W., Jr., Yin, T., *et al.* (2015). A local

VE-cadherin/Trio-based signaling complex stabilizes endothelial junctions through Rac1. J Cell Sci *128*, 3041–3054.

Tinoco, R., Otero, D.C., Takahashi, A.A., and Bradley, L.M. (2017). PSGL-1: a new player in the immune checkpoint landscape. Trends Immunol. *38*, 323–335.

Tseng, Y., Kole, T.P., Lee, J.S., Fedorov, E., Almo, S.C., Schafer, B.W., and Wirtz, D. (2005). How actin crosslinking and bundling proteins cooperate to generate an enhanced cell mechanical response. Biochem. Biophys. Res. Commun. *334*, 183–192.

van Buul, J.D., Allingham, M.J., Samson, T., Meller, J., Boulter, E., Garcia-Mata, R., and Burridge, K. (2007a). RhoG regulates endothelial apical cup assembly downstream from ICAM1 engagement and is involved in leukocyte trans-endothelial migration. J. Cell Biol. *178*, 1279–1293.

van Buul, J.D., Kanters, E., and Hordijk, P.L. (2007b). Endothelial signaling by Ig-like cell adhesion molecules. Arterioscler. Thromb. Vasc. Biol. *27*, 1870–1876.

van Buul, J.D., van Rijssel, J., van Alphen, F.P., Hoogenboezem, M., Tol, S., Hoeben, K.A., van Marle, J., Mul, E.P., and Hordijk, P.L. (2010a). Inside-out regulation of ICAM-1 dynamics in TNF-α-activated endothelium. PLoS ONE *5*, e11336.

van Buul, J.D., van Rijssel, J., van Alphen, F.P., van Stalborch, A.M., Mul, E.P., and Hordijk, P.L. (2010b). ICAM-1 clustering on endothelial cells recruits VCAM-1. J. Biomed. Biotechnol. *2010*, 120328.

van Buul, J.D., Voermans, C., van den Berg, V., Anthony, E.C., Mul, F.P., van Wetering, S., van der Schoot, C.E., and Hordijk, P.L. (2002). Migration of human hematopoietic progenitor cells across bone marrow endothelium is regulated by vascular endothelial cadherin. J. Immunol. *168*, 588–596.

van Nieuw Amerongen, G.P., Draijer, R., Vermeer, M.A., and van Hinsbergh, V.W. (1998). Transient and prolonged increase in endothelial permeability induced by histamine and thrombin: role of protein kinases, calcium, and RhoA. Circ. Res. *83*, 1115–1123.

van Nieuw Amerongen, G.P., van Delft, S., Vermeer, M.A., Collard, J.G., and van Hinsbergh, V.W. (2000). Activation of RhoA by thrombin in endothelial hyperpermeability: role of Rho kinase and protein tyrosine kinases. Circ. Res. *87*, 335–340.

van Rijssel, J., Hoogenboezem, M., Wester, L., Hordijk, P.L., and van Buul, J.D. (2012a). The N-terminal DH-PH domain of Trio induces cell spreading and migration by regulating lamellipodia dynamics in a Rac1-dependent fashion. PLoS ONE *7*, e29912.

van Rijssel, J., Kroon, J., Hoogenboezem, M., van Alphen, F.P., de Jong, R.J., Kostadinova, E., Geerts, D., Hordijk, P.L., and van Buul, J.D. (2012b). The Rho-guanine nucleotide exchange factor Trio controls leukocyte transendothelial migration by promoting docking structure formation. Mol. Biol. Cell *23*, 2831–2844.

van Rijssel, J., Timmerman, I., van Alphen, F.P.J., Hoogenboezem, M., Korchynskyi, O., Geerts, D., Geissler, J., Reedquist, K.A., Niessen, H.W.M., and van Buul, J.D. (2013). The Rho-GEF Trio regulates a novel pro-inflammatory pathway through the transcription factor Ets2. Biol. Open *2*, 569–579.

van Wetering, S., van den Berk, N., van Buul, J.D., Mul, F.P., Lommerse, I., Mous, R., ten Klooster, J.P., Zwaginga, J.J., and Hordijk, P.L. (2003). VCAM-1-mediated Rac signaling controls endothelial cell–cell contacts and leukocyte transmigration. Am. J. Physiol. Cell Physiol. *285*, C343–C352.

Vestweber, D. (2015). How leukocytes cross the vascular endothelium. Nat. Rev. Immunol. *15*, 692–704.

Vestweber, D., Zeuschner, D., Rottner, K., and Schnoor, M. (2013). Cortactin regulates the activity of small GTPases and ICAM-1 clustering in endothelium: implications for the formation of docking structures. Tissue Barriers *1*, e23862.

Vicente-Manzanares, M., Ma, X., Adelstein, R.S., and Horwitz, A.R. (2009). Non-muscle myosin II takes centre stage in cell adhesion and migration. Nat. Rev. Mol. Cell Biol. *10*, 778–790.

Vischer, U.M., Jornot, L., Wollheim, C.B., and Theler, J.M. (1995). Reactive oxygen intermediates induce regulated secretion of von Willebrand factor from cultured human vascular endothelial cells. Blood *85*, 3164–3172.

Vockel, M. and Vestweber, D. (2013). How T cells trigger the dissociation of the endothelial receptor phosphatase VE-PTP from VE-cadherin. Blood *122*, 2512–2522.

Wennerberg, K. and Der, C.J. (2004). Rho-family GTPases: it's not only Rac and Rho (and I like it). J. Cell Sci. *117*, 1301–1312.

Whittall, C., Kehoe, O., King, S., Rot, A., Patterson, A., and Middleton, J. (2013). A chemokine self-presentation mechanism involving formation of endothelial surface microstructures. J. Immunol. *190*, 1725–1736.

Wojciak-Stothard, B., Williams, L., and Ridley, A.J. (1999). Monocyte adhesion and spreading on human endothelial cells is dependent on Rho-regulated receptor clustering. J. Cell Biol. *145*, 1293–1307.

Yang, S.X., Yan, J., Deshpande, S.S., Irani, K., and Lowenstein, C.J. (2004). Rac1 regulates the release of Weibel-Palade Bodies in human aortic endothelial cells. Chin. Med. J. 117, 1143–1150.

Index